S343 Block 3

Science: a third level course

S343 INORGANIC CHEMISTRY

TRANSITION-METAL CHEMISTRY: THE STABILITIES OF OXIDATION STATES

Prepared by an Open University Course Team

S343 Course Team

Course team chairs *Stuart Bennett*

Authors
Block 1 *David Johnson*
Block 2 *Rob Janes and Elaine Moore*
Block 3 *Kiki Warr, with contributions from David Johnson*
Block 4 *Stuart Bennett*
Block 5 *Michael Mortimer*
Block 6 *Ivan Parkin, with contributions from Dr M. Kilner, Professor K. Wade* (University of Durham) *and F. R. Hartley* (Royal Military College of Science, Shrivenham)
Block 7 *Paul Walton* (University of York)*, with contributions from Lesley Smart*
Block 8 *Lesley Smart, with contributions from David Johnson, Kiki Warr and Elaine Moore*
Block 9 *David Johnson*

Consultants *Dr P. Baker* (University College of North Wales)
Dr R. Murray (Trent Polytechnic)

Course managers *Peter Fearnley*
Wendy Selina
Charlotte Sweeney

Editors *Ian Nuttall*
David Tillotson

BBC *Andrew Crilly*
David Jackson
Jack Koumi
Michael Peet

Graphic artists *Steve Best*
Janis Gilbert

Graphic designer *Josephine Cotter*

Assistance was also received from the following people: *George Loveday* (Staff Tutor), *Joan Mason, Jane Nelson* (Staff Tutor)

Course assessor *Professor J. F. Nixon* (University of Sussex)

This publication forms part of the Open University course S343 Inorganic Chemistry.The complete list of texts which make up this course can be found at the back. Details of this and other Open University courses can be obtained from the Course Information and Advice Centre, PO Box 724, The Open University, Milton Keynes MK7 6ZS, United Kingdom: tel. +44 (0)1908 653231, e-mail general-enquiries@open.ac.uk

Alternatively, you may visit the Open University website at http://www.open.ac.uk where you can learn more about the wide range of courses and packs offered at all levels by The Open University.

To purchase a selection of Open University course materials visit the webshop at www.ouw.co.uk, or contact Open University Worldwide, Michael Young Building, Walton Hall, Milton Keynes MK7 6AA, United Kingdom for a brochure. tel. +44 (0)1908 858785; fax +44 (0)1908 858787; e-mail ouwenq@open.ac.uk

The Open University
Walton Hall, Milton Keynes
MK7 6AA

First published 1989. Second edition 2004.

Edited, designed and typeset by The Open University.

Printed and bound in the United Kingdom by Henry Ling Limited, at the Dorset Press, Dorchester, DT1 1HD.

ISBN 0 7492 6637 6

2.1

BLOCK 3
TRANSITION-METAL CHEMISTRY: THE STABILITIES OF OXIDATION STATES

Prepared by an Open University Course Team
The Open University

STUDY GUIDE FOR BLOCK 3

This Block has two components—the main text and an audiocassette sequence: it should take a week of study time. There is no *new* videocassette material associated with the Block, but it does refer to experiments demonstrated in the short sequences for Block 1.

The Block comprises two main parts, each largely concerned with the relative stabilities of the +2 and +3 oxidation states of metals of the first transition series. The first part runs from the Introduction to the end of Section 3.1. It uses thermodynamics and the theories introduced in Block 2, to complete our analysis of the cross-series stability problem raised in Block 1. The audiocassette sequence is associated with part of this material (Section 2.3).

The second part of the Block (Sections 4 to 7) looks at a complementary problem: the effect of complex formation on the relative stability of oxidation states of a given transition element. To analyse this problem requires two results of far-reaching importance—the stability constant of a complex and the Nernst equation: these are introduced in Sections 4.2 and 5.2, respectively. It is particularly important that you are confident about using the Nernst equation—and, indeed, electrode potentials in general—so it would be wise to tackle as many as possible of the SAQs in Section 5.

1 INTRODUCTION

We began the Course by giving you a taste of the richness and variety of transition-metal chemistry, a variety that reflects the extraordinary range of oxidation states displayed by these elements. Armed with the theoretical treatment developed in the last Block, we are now in a position to take a closer look at certain aspects of this chemistry. We shall concentrate on two main areas.

The first is a problem that was raised in Block 1: the variation in the relative stabilities of oxidation states +2 and +3 across the first transition series. Here we embark on a detailed analysis of this problem, one that draws on your newly acquired understanding of ligand-field stabilisation energies, and their consequences.

Our second major topic is again concerned with variations in the relative stability of different oxidation states. But, whereas the patterns unearthed in Block 1 relate to systems like the following:

$$M^{2+}(aq) + H^{+}(aq) = M^{3+}(aq) + \tfrac{1}{2}H_2(g) \qquad 1$$

where the ligands are fixed and the transition element is varied, we turn at this point to the complementary case—to problems of relative stability in which the transition element is fixed and the ligand is changed. Examples can again be drawn from Block 1, a case in point being the stabilisation of cobalt(III) by complexing (with ammonia, NH_3, for instance) or by the formation of insoluble compounds such as $Co(OH)_3$.

To discuss problems like this in a quantitative way, we shall need to add two important new 'weapons' to our thermodynamic 'armoury'. These are the idea of a stability constant of a transition-metal complex, and the Nernst equation. As you will see (both here and later on in the Course), taken together, these ideas give us a handle on some fascinating aspects of transition-metal chemistry, aspects that have consequences in areas as diverse as the problem of supplying sufficient micronutrients to plants, or understanding the subtle ligand effects that modulate the action of many metal-containing enzymes.

To return to our first problem, let us start with a brief reminder of the experimental observations collected in Block 1 (Sections 8 and 9), and the conclusions drawn from them.

2 RELATIVE STABILITIES OF DIPOSITIVE AND TRIPOSITIVE AQUEOUS IONS OF THE FIRST TRANSITION SERIES

2.1 Introduction

Every metal of the first transition series *except* scandium and titanium forms an aqueous dipositive ion, whereas the corresponding tripositive ions are known from scandium to cobalt, but not beyond. In other words, the dipositive ions at the beginning of the series, Sc^{2+}(aq) and Ti^{2+}(aq), are unknown; and the tripositive ions at the end, Ni^{3+}(aq), Cu^{3+}(aq) and Zn^{3+}(aq), are unknown.

This pattern suggests a trend, which can be described in two alternative ways. On the one hand, it apparently gets harder to oxidise M^{2+}(aq) to M^{3+}(aq) across the series. Alternatively, it seems that M^{3+}(aq) gradually becomes a more powerful oxidising agent. In the main, this trend is corroborated by the behaviour of the elements in the middle of the series, in the vanadium to cobalt region. Thus, acid solutions of V^{2+}(aq) and Cr^{2+}(aq) are oxidised by atmospheric oxygen, but Co^{3+}(aq) gradually oxidises water, thereby *producing* oxygen. Equally, Co^{3+}(aq) will oxidise Mn^{2+}(aq) to Mn^{3+}(aq) and Fe^{2+}(aq) to Fe^{3+}(aq), and hence is a stronger oxidising agent than either Mn^{3+}(aq) or Fe^{3+}(aq).

But recall that there is one important point where the trend is broken: rather than Fe^{3+}(aq) being a stronger oxidising agent than Mn^{3+}(aq), the converse is true. Thus, it is more accurate to say that *the stability of M^{2+}(aq), with respect to oxidation to M^{3+}(aq), shows an overall increase across the series, but there is an important decrease from manganese to iron.*

As you have seen, this trend can be quantified in terms of the relative thermodynamic stability of the two ions, as expressed by values of $E^{\ominus}(M^{3+}|M^{2+})$, or alternatively by values of $\Delta G_m^{\ominus}$ for the process in equation 1:

$$M^{2+}(aq) + H^{+}(aq) = M^{3+}(aq) + \tfrac{1}{2}H_2(g) \qquad 1$$

the larger (more positive) the value of $\Delta G_m^{\ominus}(1)$, the more stable is the M^{2+}(aq) ion. As shown by the lower plot in Figure 1, there is indeed an overall increase in the stability of M^{2+}(aq) across the series, with a significant downward break from manganese to iron.

Our final step in Block 1 was to point out that variations in $\Delta G_m^{\ominus}(1)$ record variations in the difficulty of removing an electron from an *aqueous* M^{2+} ion. It came as no real surprise to find that these variations parallel those in the difficulty of removing an electron from a *gaseous* M^{2+} ion, as recorded by the third ionisation energies, I_3, of the metals—effectively the standard molar enthalpy change, $\Delta H_m^{\ominus}$, for the process:

$$M^{2+}(g) = M^{3+}(g) + e^{-} \qquad 2$$

The comparison is repeated here as Figure 1: in both plots there is an overall increase across the series and a distinct downward break at the halfway point, between manganese and iron.

As we noted at the time, however, the parallelism is far from perfect: the overall variation in $\Delta G_m^{\ominus}(1)$ looks like a watered-down version of that in I_3—and it also contains one or two extra, if small, irregularities. For example, it is noticeable that from vanadium to chromium, $\Delta G_m^{\ominus}(1)$ changes in the opposite sense to I_3.

So much for our brief review of the stability problem as stated in Block 1. To probe more deeply into the problem, our first task is to tighten up the arguments rehearsed above, which are based on a rather loose analogy between 'the ease of removal of an electron' from M^{2+}(g) and M^{2+}(aq). To this end, we begin the next Section by using thermodynamics to establish a precise relationship between $\Delta G_m^{\ominus}(1)$ and I_3.

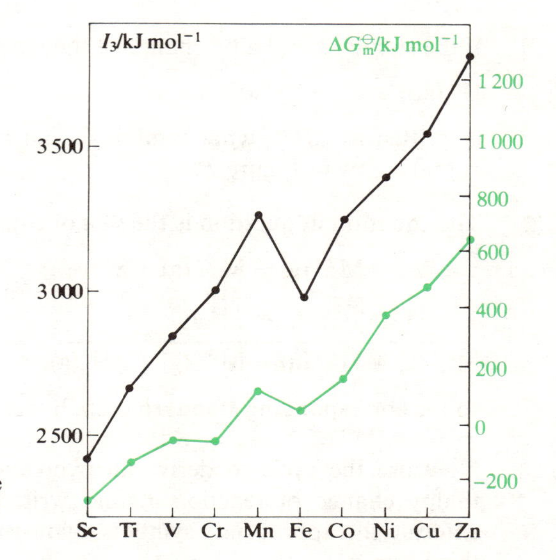

Figure 1 Variations in $\Delta G^{\ominus}_{m}$ for reaction 1 (green line) and I_3 (black line) across the first transition series. The plots are on the same scale, but with different origins.

Having forged this link, we shall then be in a position to examine the factors that cause departures from parallelism between $\Delta G^{\ominus}_{m}(1)$ and I_3. And having done that, we shall find that our original objective—to explain the cross-series variation in $\Delta G^{\ominus}_{m}$—largely resolves itself into the problem of explaining the variation in I_3.

2.2 Analysis by a thermodynamic cycle

I_3 is virtually identical with $\Delta H^{\ominus}_{m}$ for the process in equation 2, and here we shall assume that they are equal. This suggests that a good starting point is to construct a thermodynamic cycle that links equations 1 and 2, and then to consider the associated standard enthalpy terms. A suitable cycle is shown in Figure 2.

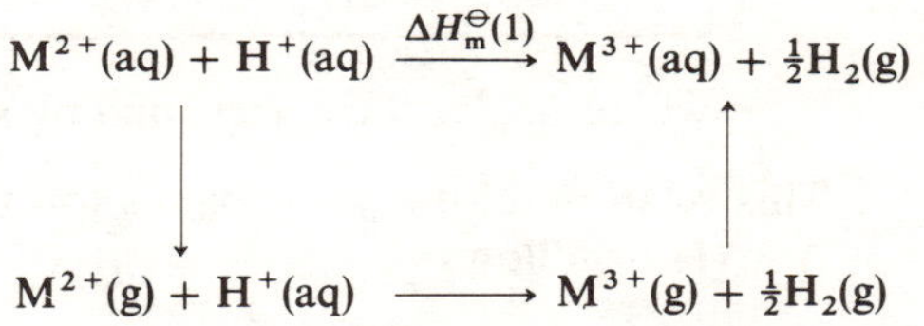

$$M^{2+}(aq) + H^{+}(aq) \xrightarrow{\Delta H^{\ominus}_{m}(1)} M^{3+}(aq) + \tfrac{1}{2}H_2(g) \qquad 1$$

$$\downarrow \qquad\qquad\qquad\qquad \uparrow$$

$$M^{2+}(g) + H^{+}(aq) \longrightarrow M^{3+}(g) + \tfrac{1}{2}H_2(g)$$

Figure 2 Thermodynamic cycle around reaction 1.

☐ If the standard molar enthalpy of hydration of the ion $M^{n+}(g)$ is written as $\Delta H^{\ominus}_{h}(M^{n+}, g)$, what symbol should be written against the vertical arrow on the right-hand side of Figure 2?

■ $\Delta H^{\ominus}_{h}(M^{3+}, g)$. The right-hand arrow represents the equation

$$M^{3+}(g) + \tfrac{1}{2}H_2(g) = M^{3+}(aq) + \tfrac{1}{2}H_2(g)$$

which reduces to the 'hydration reaction' for M^{3+} when the common term $\tfrac{1}{2}H_2(g)$ is eliminated from both sides, that is:

$$M^{3+}(g) = M^{3+}(aq) \qquad 3$$

Similar reasoning shows that the appropriate symbol to place against the left-hand arrow is $-\Delta H^{\ominus}_{h}(M^{2+}, g)$. Here, the minus sign arises because the arrow points *from* the aqueous ion *to* the gaseous ion, indicating that the process in question is the *reverse* of the hydration reaction for $M^{2+}(g)$, that is:

$$M^{2+}(aq) + H^{+}(aq) = M^{2+}(g) + H^{+}(aq)$$

☐ If the standard enthalpy change of the reaction:

$$H^+(aq) + e^- = \tfrac{1}{2}H_2(g) \quad \mathbf{4}$$

is written as $\Delta H^\ominus_H$, what symbol should be written against the lower horizontal arrow in Figure 2?

■ The equation in question is the *sum* of equations 2 and 4, that is:

$$M^{2+}(g) = M^{3+}(g) + e^- \quad \mathbf{2}$$

$$H^+(aq) + e^- = \tfrac{1}{2}H_2(g) \quad \mathbf{4}$$

$$M^{2+}(g) + H^+(aq) = M^{3+}(g) + \tfrac{1}{2}H_2(g) \quad \mathbf{5}$$

so the corresponding standard enthalpy change is just $(I_3 + \Delta H^\ominus_H)$.

☐ Now use the cycle to derive an expression for $\Delta H^\ominus_m(1)$, the standard enthalpy change of reaction 1, and write it in the space below. Start by entering the appropriate symbols against the arrows in Figure 2 if you find this helps.

■

SLC1

☐ Do you recall the relation between $\Delta G^\ominus_m$ and $\Delta H^\ominus_m$, as introduced in a Second Level Course?

■ This important relation comes from the *definition of the Gibbs function* ($G = H - TS$), and is given, for a reaction at constant temperature T by the expression:

$$\Delta G^\ominus_m = \Delta H^\ominus_m - T\Delta S^\ominus_m \quad \mathbf{6}$$

where $\Delta S^\ominus_m$ is the corresponding *standard molar entropy change.*

This relation allows your expression for $\Delta H^\ominus_m(1)$ to be transformed into one for $\Delta G^\ominus_m(1)$, as follows:

$$\Delta G^\ominus_m(1) = -\Delta H^\ominus_h(M^{2+}, g) + (I_3 + \Delta H^\ominus_H) + \Delta H^\ominus_h(M^{3+}, g) - T\Delta S^\ominus_m(1) \quad \mathbf{7}$$

Check that your expression for $\Delta H^\ominus_m(1)$ is the right-hand side of this equation without the term $-T\Delta S^\ominus_m(1)$, where $\Delta S^\ominus_m(1)$ again refers to reaction 1.

SLC 2

Now, we are interested only in *variations* in the quantities in equation 7 across the transition series, so $\Delta H^\ominus_H$—which is constant—can be ignored. Moreover, arguments advanced in a Second Level Course suggest that $\Delta S^\ominus_m$, and hence $T\Delta S^\ominus_m$, is likely to change relatively little for a series of reactions as closely analogous as those under consideration here. Thus, across the first transition series, $\Delta G^\ominus_m(1)$ can be approximated by a simplified version of equation 7, that is:

$$\Delta G^\ominus_m(1) \approx I_3 + \Delta H^\ominus_h(M^{3+}, g) - \Delta H^\ominus_h(M^{2+}, g) + C \quad \mathbf{8}$$

where C is a constant.

Equation 8 is the desired relationship between $\Delta G^\ominus_m(1)$ and I_3, and it serves to confirm that the variation in $\Delta G^\ominus_m(1)$ should indeed be influenced by that in I_3. But we can now take the analysis a step further. The experimental results in Figure 1 certainly suggest that the form of the variation in I_3 is the dominant effect, but it is not the only one. Obviously, the two other variables on the right-hand side of equation 8 also exert an influence; otherwise the parallelism between $\Delta G^\ominus_m(1)$ and I_3 would be perfect. In fact, the distinct differences between the variations in $\Delta G^\ominus_m(1)$ and I_3 must be due to irregular variations in $\Delta H^\ominus_h(M^{2+}, g)$ and $\Delta H^\ominus_h(M^{3+}, g)$.

Ligand-field stabilisation energies (LFSEs for short), and their variations across a transition series, are central to the discussion of these irregularities. If you are at all uncertain about these ideas, check your understanding by working through Exercise 1. The answer is on p. 43.

Exercise 1 (*revision*)

(a) Experiment shows that the aqueous M^{2+} ions of the first transition series are all high-spin octahedral complexes, $[M(H_2O)_6]^{2+}(aq)$. Thus, the hydration enthalpy, $\Delta H_h^{\ominus}(M^{2+}, g)$, refers to a process in which a gaseous ion enters into octahedral coordination:

$$M^{2+}(g) = [M(H_2O)_6]^{2+}(aq)$$

(as does the lattice energy of the chlorides, $L_0(MCl_2, s)$, discussed in Block 2). With this in mind, draw a sketch to show how you would expect $\Delta H_h^{\ominus}(M^{2+}, g)$ to vary across the series from Ca^{2+} to Zn^{2+}. Explain *briefly* the main features of your sketch.

(b) The aqueous M^{3+} ions are also high-spin octahedral complexes, with the exception of $[Co(H_2O)_6]^{3+}$, which is low-spin d^6. However, this solitary exception can be ignored here because the high-spin and low-spin states are known to be close in energy. Outline *briefly* how you would expect the variation in $\Delta H_h^{\ominus}(M^{3+}, g)$ across the series to differ from the variation in $\Delta H_h^{\ominus}(M^{2+}, g)$.

To summarise: ligand-field theory suggests that the variations in both $\Delta H_h^{\ominus}(M^{2+}, g)$ and $\Delta H_h^{\ominus}(M^{3+}, g)$ should consist of a smoothly varying contribution from the d^0, d^5, d^{10} curve—together with an irregularly varying contribution from the LFSEs.

This pattern is apparent in the information collected in Figure 3. Here, the four variable quantities in equation 8 are plotted on the same scale, but the energy zeros (which are different for each term) have been adjusted in order to bring the plots together. In addition, the quantities have been plotted with the *signs* that they have in equation 8 (which turns the more familiar bowls of $+\Delta H_h^{\ominus}(M^{2+}, g)$ into a double hump with a cusp at d^5 for $-\Delta H_h^{\ominus}(M^{2+}, g)$). This means that the $\Delta G_m^{\ominus}$ plot is the simple *sum* of the others, and can be obtained by superimposing them at the beginning of the series. Finally, notice that we have also included the 'constant' $-T\Delta S_m^{\ominus}(1)$ term. This serves to confirm our earlier assumption that it contributes little to the variation in $\Delta G_m^{\ominus}$: we shall say no more about it.

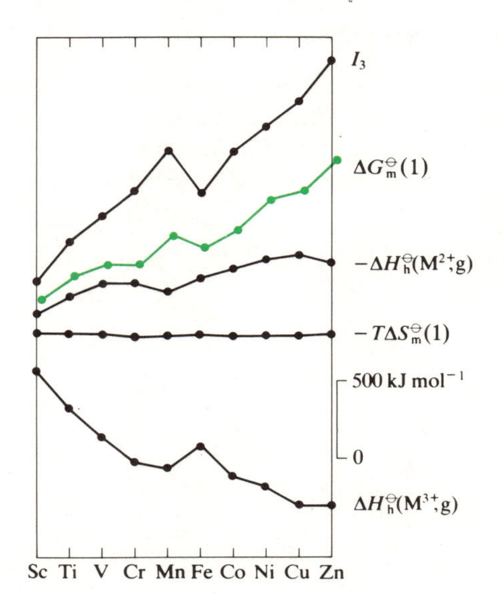

Figure 3 Variations in the quantities in equation 8 across the first transition series. Some of the data in this Figure—notably for unknown or unstable species—are estimated values.

The comparison in Figure 3 highlights the dominance of the I_3 term in equation 8. Notice, for example, that the changes in $\Delta H_h^{\ominus}(M^{3+}, g)$ are such that, taken alone, they would produce almost the exact *opposite* of the observed variation in $\Delta G_m^{\ominus}(1)$. Moreover, their adverse contribution is only partially offset by the smaller fluctuations in $-\Delta H_h^{\ominus}(M^{2+}, g)$. Nevertheless, the combined effect of these terms, $\{\Delta H_h^{\ominus}(M^{3+}, g) - \Delta H_h^{\ominus}(M^{2+}, g)\}$, is almost everywhere overwhelmed by the comparatively large overall increase in I_3: whence the observed parallelism between I_3 and $\Delta G_m^{\ominus}$.

SAQ 1 As mentioned earlier, the most marked departure from parallelism occurs from vanadium to chromium. $Cr^{3+}(aq)$ is more stable to reduction than is $V^{3+}(aq)$, but the opposite holds for the free ions in the gas phase. What term in Figure 3 is mainly responsible for this departure?

It is worth stressing that the example in SAQ 1 is the only point (in the first transition series) where the other terms in equation 8 dominate the variation in $\Delta G_m^{\ominus}$ for equation 1 at the expense of I_3. Thus, we are led to the final, and most crucial, step in our analysis—explaining the variation in I_3.

2.3 The third ionisation energy of the elements

Most of this Section takes the form of an AV sequence. We begin by noting that the gaseous M^{2+} ions of the first transition series all have electronic configurations of the type $[Ar]3d^n$. Thus, the third ionisation energy can be equated with the energy change for the process:

$$[Ar]3d^n \longrightarrow [Ar]3d^{n-1} + e^- \qquad (9)$$

as $$I_3 = E(d^{n-1}) - E(d^n) \qquad (10)$$

where $E(d^n)$ and $E(d^{n-1})$ are the energies of the $3d^n$ and $3d^{n-1}$ configurations, respectively. Hence our task becomes the calculation of the energy of the 3d shell as a function of the number of electrons n.

You should now turn to Section 5 in the S343 Audiovision Booklet, and work through the associated audiocassette sequence. The main results of our analysis are summarised below.

According to our analysis, the third ionisation energy is given by the following (albeit approximate) equation:

$$I_3 = E(d^{n-1}) - E(d^n) = U - (n-1)J + K\,\delta m \qquad (11)$$

1 $-U$ represents the energy of a 3d electron due to the coulombic attraction of the positive argon core: across the series, U increases smoothly with n as the increasing nuclear charge holds the ionising electron more tightly.

2 The second term in equation 11 stems from the classical coulombic repulsion between electrons, which is approximately proportional (through the constant J) to the number of pairs of electrons: $-(n-1)J$ is a measure of the extent to which the ionising electron is repelled by the remaining $(n-1)$.

3 The experimental values of I_3 show that $\{U - (n-1)J\}$ increases smoothly with n, the increasing nuclear charge outweighing the increasing repulsion between the d electrons: this is represented by the curve ABC in Figure 4.

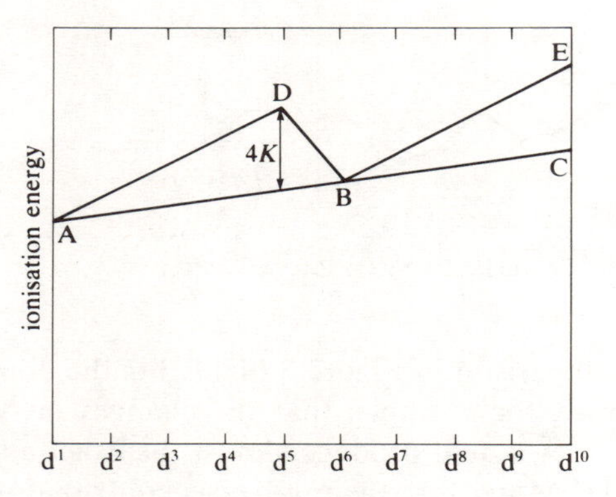

Figure 4 Schematic variation in the ionisation energies of d^n configurations.

4 The final term in equation 11 stems from the non-classical part of the repulsion energy—the **exchange energy**—which is roughly proportional (through the constant K) to the number of pairs of *parallel-spin* interactions: values of δm—the *decrease* in the number of pairs of such interactions—are collected in Table 1.

Table 1 The decrease in the number of pairs of parallel spins on ionisation for electronic configurations d^1 to d^{10}

Ionisation process	δm
$d^1 \rightarrow d^0$	0
$d^2 \rightarrow d^1$	1
$d^3 \rightarrow d^2$	2
$d^4 \rightarrow d^3$	3
$d^5 \rightarrow d^4$	4
$d^6 \rightarrow d^5$	0
$d^7 \rightarrow d^6$	1
$d^8 \rightarrow d^7$	2
$d^9 \rightarrow d^8$	3
$d^{10} \rightarrow d^9$	4

5 When the term $K\,\delta m$ is added to ABC, the plot ADBE results, which is very similar in shape to the experimental variation in I_3. When ionisation occurs, parallel-spin interactions are usually (*but not always*) destroyed, and some stabilising exchange energy is lost with them. However, this loss does not vary smoothly with n, so the variation in I_3 is irregular.

To summarise: the overall increase of the third ionisation energy is the result of inadequate compensation of the increasing nuclear charge by the repulsive or screening effect of the increasing number of d electrons. The crucial drop from manganese to iron is a consequence of the unchanged exchange energy when an electron is lost from a d^6 ion. From configurations d^1 to d^5, the loss of an electron decreases the number of pairs of parallel spins by zero, one, two, three and four, respectively, but at d^6 this decrease falls to zero again, and the d^1 to d^5 pattern is repeated from d^6 to d^{10}.

SAQ 2 Figure 5 shows the variation in the first ionisation energies of the elements from boron to neon. Use the theory developed in the AV sequence for this Section to explain it. Specify the nature of the core carefully, and include in your explanation a table corresponding to Table 1.

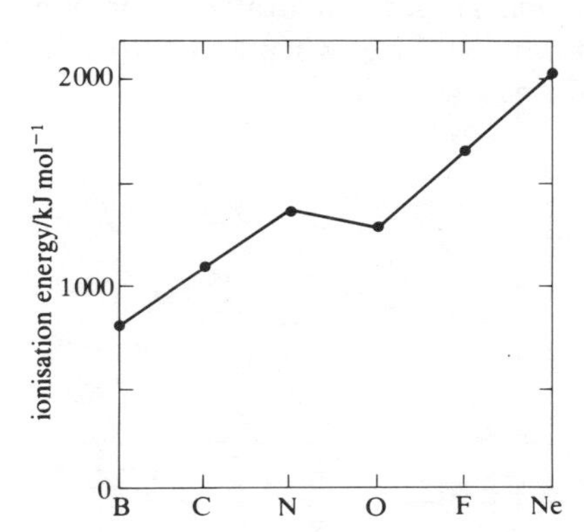

Figure 5 First ionisation energies of the elements from boron to neon.

2.4 Summary of Section 2

1 Experimental values of $E^{\ominus}(M^{3+}(aq)|M^{2+}(aq))$ suggest that in crossing the first transition series, $\Delta G^{\ominus}_{m}$ for the reaction:

$$M^{2+}(aq) + H^{+}(aq) = M^{3+}(aq) + \tfrac{1}{2}H_2(g) \qquad \mathbf{1}$$

shows an overall increase, broken by a decrease from manganese to iron. This variation in $\Delta G^{\ominus}_{m}$ parallels, albeit imperfectly, the variation in the third ionisation energies of the metals.

2 For the first-row transition elements:

$$\Delta G^{\ominus}_{m}(1) \approx I_3 + \Delta H^{\ominus}_{h}(M^{3+}, g) - \Delta H^{\ominus}_{h}(M^{2+}, g) + C \qquad \mathbf{8}$$

3 The observed parallelism arises because, among the terms on the right-hand side of equation 8, the variations in I_3 are more dramatic than the combined variations of the remaining terms.

4 The parallelism is imperfect mainly because of the irregular variations in $\Delta H^{\ominus}_{h}(M^{2+}, g)$ and $\Delta H^{\ominus}_{h}(M^{3+}, g)$, irregularities that stem from the contributions of ligand-field stabilisation energies to these two quantities.

5 The overall increase in I_3 is the result of inadequate compensation of the increasing nuclear charge by the repulsive or screening effect of the increasing number of d electrons. The crucial drop from manganese to iron is caused by the phenomenon of an unchanged exchange energy when an electron is lost from a d^6 ion.

6 According to our analysis, this explanation of the variation in I_3 also serves to explain that in $\Delta G^{\ominus}_{m}(1)$.

□ The main impetus for the analysis summarised above was to provide a rationalising framework for the numerous experimental observations collected in Block 1 and its associated video material. Before reading further, jot down as many of these as you can remember, and try to assess how they reflect the variation in I_3 or $\Delta G_m^{\ominus}(1)$ across the series.

■ Early in the series, where I_3 is relatively low, the ion $M^{2+}(aq)$ is strongly reducing, and $Sc^{2+}(aq)$ and $Ti^{2+}(aq)$ are unknown because they reduce $H^+(aq)$ or water. Thus, $TiCl_2$ instantly evolves hydrogen when added to dilute acids. Aqueous solutions of V^{2+} and Cr^{2+} do exist, but they are still powerful reducing agents and are oxidised to $M^{3+}(aq)$ in air. However, at manganese, the increase in I_3 has made $Mn^{3+}(aq)$ a powerful oxidising agent, and even more powerful oxidising agents like persulphate are needed to convert $Mn^{2+}(aq)$ to $Mn^{3+}(aq)$. The fall in I_3 from d^5 to d^6 can be demonstrated by oxidising $Fe^{2+}(aq)$ with $Mn^{3+}(aq)$. From iron to cobalt the renewed increase in the stability of $M^{2+}(aq)$ becomes apparent because, unlike $Fe^{3+}(aq)$, $Co^{3+}(aq)$ is steadily reduced to $Co^{2+}(aq)$ by water. For nickel, copper and zinc, the $M^{3+}(aq)$ ions have become such powerful oxidising agents that they oxidise water instantly and are therefore unknown. Thus, the hydrated oxide of nickel(III), NiO(OH), evolves oxygen and forms $Ni^{2+}(aq)$ when added to acid; likewise, $K_3[CuF_6]$ oxidises water and forms $Cu^{2+}(aq)$.

3 EXTENSION TO OTHER SYSTEMS

The process (equation 1) considered at such length in the previous Section is just one example of a reaction in which a transition element is converted from oxidation state +2 to +3. To what extent can the analysis developed there be extended to other, apparently similar, systems?

A case in point is the problem of the relative stability of dihalides (MX_2) and trihalides (MX_3) across the first transition series, as introduced in Block 1. There, we argued that the range of known compounds (repeated here as Tables 2 and 3) could be seen to reflect variations in the value of $\Delta G_m^{\ominus}$ for an oxidation reaction of the general type:

$$MX_2(s) + \tfrac{1}{2}X_2 = MX_3(s) \qquad 12$$

where X is the halogen in question.

Table 2 The range of known dihalides of the first-row transition elements

	Sc	Ti	V	Cr	Mn	Fe	Co	Ni	Cu	Zn
MF_2	—	—	✓	✓	✓	✓	✓	✓	✓	✓
MCl_2	—	✓	✓	✓	✓	✓	✓	✓	✓	✓
MBr_2	—	✓	✓	✓	✓	✓	✓	✓	✓	✓
MI_2	—	✓	✓	✓	✓	✓	✓	✓	—	✓

Table 3 The range of known trihalides of the first-row transition elements

	Sc	Ti	V	Cr	Mn	Fe	Co	Ni	Cu	Zn
MF_3	✓	✓	✓	✓	✓	✓	✓	—	—	—
MCl_3	✓	✓	✓	✓	*	✓	—	—	—	—
MBr_3	✓	✓	✓	✓	—	✓	—	—	—	—
MI_3	✓	✓	✓	✓	—	—	—	—	—	—

* Decomposes above −40 °C.

Suppose we start with the chlorides. The comparison between the appropriate version of equation 12,

$$MCl_2(s) + \tfrac{1}{2}Cl_2(g) = MCl_3(s) \qquad \mathbf{13}$$

and the 'standard' reaction, equation 1, is repeated here as Figure 6. Once again, one is struck by the close similarity between the variations in $\Delta G_m^{\ominus}$ for the two reactions. And if *these* quantities are so alike, then it follows that $\Delta G_m^{\ominus}$ for the chloride reaction must also show a marked resemblance to the familiar 'saw-tooth' variation in I_3. Now work through the following Exercise.

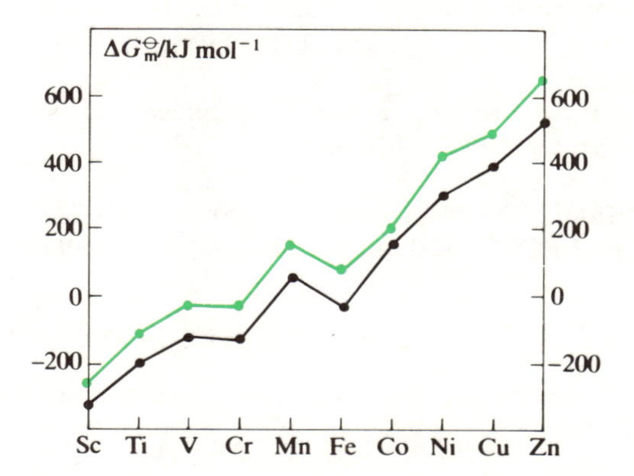

Figure 6 Variations in $\Delta G_m^{\ominus}$ for reaction 1 (green line) and reaction 13 (black line) across the first transition series.

Exercise 2

(a) Consider the thermodynamic cycle shown in Figure 7. By inserting appropriate symbols against the unlabelled enthalpy changes, establish a relationship between the variations in $\Delta G_m^{\ominus}(13)$ and I_3 across the first transition series. Write out an equation corresponding to equation 8.

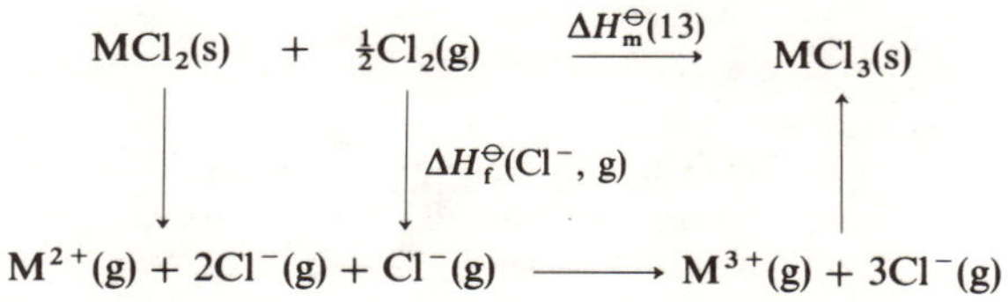

$MCl_2(s) + \tfrac{1}{2}Cl_2(g) \xrightarrow{\Delta H_m^{\ominus}(13)} MCl_3(s)$ **13**

$\Delta H_f^{\ominus}(Cl^-, g)$

$M^{2+}(g) + 2Cl^-(g) + Cl^-(g) \longrightarrow M^{3+}(g) + 3Cl^-(g)$

Figure 7 Thermodynamic cycle around reaction 13.

(b) By reference to your equation from part (a), explain *briefly* why the variation in $\Delta G_m^{\ominus}(13)$ parallels that in I_3. How would you account for the departures from parallelism?

To summarise: in answering this Exercise, you followed closely our attack on the problem analysed in Section 2. There are three stages. First (in part (a)), you established a precise relationship between $\Delta G_m^{\ominus}(13)$ and I_3. Next (in part (b)), you explained the departures from parallelism by commenting on the irregular contributions of LFSEs to the lattice energy terms of both MCl_2 and MCl_3. These mitigate the changes in I_3. The interpretation of these changes represents the third, and final stage: obviously, this would be identical with Section 2.3.

As suggested earlier, the *chemical* consequences of these variations can be inferred from Tables 2 and 3. Thus, at the beginning of the series (where I_3 is relatively low) the value of $\Delta G_m^{\ominus}(13)$ is so negative that $ScCl_2$ is unknown at room temperature. Evidently, it is so unstable with respect to $ScCl_3$ that it must disproportionate, as:

$$3ScCl_2(s) = Sc(s) + 2ScCl_3(s) \qquad \mathbf{14}$$

SAQ 3 (*revision*) Why should the low stability of $ScCl_2$ with respect to reaction 13 encourage disproportionation via the reaction in equation 14?

Moving along the series, $TiCl_2$, VCl_2 and $CrCl_2$ are stable to disproportionation. By manganese, I_3 has become so large that the trichloride is incompletely characterised and does not exist at room temperature: here, $\Delta G_m^{\ominus}(13)$ is positive (Figure 6), and $MnCl_3$ readily loses chlorine via the *reverse* of equation 13. By contrast, $FeCl_3$ is stable with respect to loss of chlorine at room temperature, that is,

$\Delta G^{\ominus}_{m}(13)$ is negative (Figure 6)—thus reflecting the decrease in I_3 between manganese and iron. But from cobalt to zinc, the renewed increase in I_3 is apparent from the total instability of the unknown trichlorides.

So much for the chlorides. But what about the other halide systems—further manifestations of the general reaction:

$$MX_2(s) + \tfrac{1}{2}X_2 = MX_3(s) \qquad \textbf{12}$$

but where X is now F, Br or I? These systems can be analysed in just the same way as the chlorides. And when they are, we find that much the same general conclusions apply: in each case, the stability sequence parallels the variation in I_3. Convince yourself of this by working through the following SAQs.

SAQ 4 To what extent is the influence of the I_3 variation manifest in the information about fluorides conveyed by Tables 2 and 3?

SAQ 5 From manganese to iron, I_3 falls. Is this apparent in the information conveyed by Tables 2 and 3 on (a) bromides, and (b) iodides?

3.1 The influence of ligand-field effects

To review the argument: so far, we have looked at five reactions:

$$M^{2+}(aq) + H^{+}(aq) = M^{3+}(aq) + \tfrac{1}{2}H_2(g) \qquad \textbf{1}$$

and $MX_2(s) + \tfrac{1}{2}X_2 = MX_3(s)$ **12**

where X = F, Cl, Br or I. In each case, we have evidence that the variation in $\Delta G^{\ominus}_{m}$ across the first transition series more-or-less mirrors that in I_3. At this point, it is tempting to assert that the same trend should hold for *all* such reactions, that is, those in which a first-row transition element is oxidised from oxidation state +2 to oxidation state +3.

To test this assertion, start by working through the following SAQ.

SAQ 6 The group of metals in the middle of the first transition series—chromium, manganese, iron and cobalt—are known to form cyanide complexes (with CN^-, as ligand) in oxidation states +2 and +3. With one exception*, the complexes are octahedral, such that the relative stability of the M^{II} and M^{III} complexes, $[M(CN)_6]^{4-}$ and $[M(CN)_6]^{3-}$, respectively, can be assessed in terms of $\Delta G^{\ominus}_{m}$ for a reaction analogous to equation 1:

$$[M(CN)_6]^{4-}(aq) + H^{+}(aq) = [M(CN)_6]^{3-}(aq) + \tfrac{1}{2}H_2(g) \qquad \textbf{15}$$

Standard redox potentials for the cyanide couples are collected in Table 4. According to these data, how does $\Delta G^{\ominus}_{m}$ for the reaction in equation 15 vary along the sequence Cr, Mn, Fe, Co? Does this variation depart noticeably from that in I_3?

Table 4 Standard electrode potentials at 298.15 K for selected couples of the type $[M^{III}(CN)_6]^{3-}/[M^{II}(CN)_6]^{4-}$

Couple	$E^{\ominus}$/V
$[Cr(CN)_6]^{3-}(aq) + e = [Cr(CN)_6]^{4-}(aq)$	−1.14
$[Mn(CN)_6]^{3-}(aq) + e = [Mn(CN)_6]^{4-}(aq)$	−0.24
$[Fe(CN)_6]^{3-}(aq) + e = [Fe(CN)_6]^{4-}(aq)$	+0.36
$[Co(CN)_6]^{3-}(aq) + e = [Co(CN)_6]^{4-}(aq)$	−1.0

The reason for this marked departure from a stability series dominated by changes in I_3 is not hard to find. It can be traced to the fact that the five ligands involved in reactions 1 and 12 have something in common—something that they do *not* share with cyanide, CN^-.

☐ Where do the five ligands—the four halides and water—occur in the spectrochemical series? What about CN^-?

* The exception is the cyanocobalt(II) complex: this is green in water and has the formula $[Co(CN)_5]^{3-}(aq)$ *not* $[Co(CN)_6]^{4-}(aq)$. The argument here assumes that it makes no difference to regard it as the latter.

■ The first five all occur at the weak-field end, whereas cyanide is most definitely a strong-field ligand.

But how does this explain the somewhat different stability pattern of the cyanide system, equation 15? Well, recall (from Section 2.2 and Exercise 2) that departures from parallelism between $\Delta G_m^\ominus$ and I_3 may be largely attributed to irregular contributions from the LFSEs. Weak-field ligands produce relatively small ligand-field splittings, Δ_o, which is why the octahedrally coordinated compounds and complexes in our first five reactions are all (with one sole exception, see Exercise 1) high spin. The crucial point is that the bowls in lattice energy variations or in enthalpies of hydration are also relatively small—and hence are unable to make a decisive impression on the more dramatic changes in I_3.

Strong-field ligands, like CN^-, provide a strikingly different situation. Now some, or even all, of the d^4, d^5, d^6 and d^7 complexes will be *low spin*—and this is so for the cyanide complexes in Table 4. The stability problem may, of course, be restated by way of a thermodynamic cycle, analogous to that in Figure 2 (Section 2.2). It then becomes clear that very large ligand-field stabilisation energies (dependent not only on Δ_o, but also on the *pairing energy* for low-spin complexes) will cause far more dramatic variations in the metal–ligand interaction terms (that is, the enthalpy changes analogous to $\Delta H_h^\ominus$ or L_0 for reactions 1 and 12, respectively). A detailed analysis of the cyanide system suggests that this is indeed what destroys the parallelism between $\Delta G_m^\ominus$ and I_3.

To summarise: the cyanide example shows how ligand-field stabilisation energies can, if sufficiently large, break the useful correlation that seemed conceivable earlier in this Section. As a result, I_3 dominates the variation in the relative stabilities of the +2 and +3 oxidation states only for complexes with weak-field ligands. Other, rather unpredictable, variations occur with strong-field ligands: in consequence, there is no longer one particular and unique pattern of relative stabilities across the first transition series.

4 COMPLEX FORMATION AND THE STABILITIES OF OXIDATION STATES—SIGNPOSTING THE WAY AHEAD

4.1 Introduction

So far, we have been taken up with one particular type of problem: the variation in the relative stability of two oxidation states when the ligands are fixed and the transition element is varied. Throughout most of this discussion, the *nature* of the ligand has played a somewhat subsidiary role. From now on, we shall focus on this role, and enquire about the effect that changing the ligands has on the relative stability of different oxidation states of a *given* transition element. Although we did not stress this aspect at the time, the cobalt chemistry discussed in Block 1 (and demonstrated in the associated video material) provides a particularly dramatic example.

Let's review the salient points. $Co^{3+}(aq)$ is a powerful oxidising agent. Thus, oxidation of $Co^{2+}(aq)$ is very difficult, but possible: electrolysis in acid solution at 0 °C does the trick, thereby producing the deep blue colour of $Co^{3+}(aq)$. But on standing at room temperature (or warming), $Co^{3+}(aq)$ steadily oxidises water: oxygen is evolved, and the blue colour of $Co^{3+}(aq)$ disappears, to be replaced by the pink of $Co^{2+}(aq)$.

□ Are these observations consistent with the following standard electrode potentials?

$$Co^{3+}(aq) + e = Co^{2+}(aq); \quad E^\ominus = 1.94\,V \qquad \mathbf{16}$$

$$\tfrac{1}{2}O_2(g) + 2H^+(aq) + 2e = H_2O(l); \quad E^\ominus = 1.23\,V \qquad \mathbf{17}$$

■ Yes. $E^{\ominus}(Co^{3+} | Co^{2+})$ is greater (more positive) than $E^{\ominus}$ for the oxygen/water couple. Thus, in acid solution at 25 °C, $Co^{3+}(aq)$ is thermodynamically capable of oxidising water.

Put another way, electrode potentials tell us that the system represented by the following equation:

$$2Co^{3+}(aq) + H_2O(l) = 2Co^{2+}(aq) + 2H^+(aq) + \tfrac{1}{2}O_2(g) \quad \textbf{18}$$

has $\Delta G_m^{\ominus} < 0$, so equilibrium lies over to the right.

Consider now the effect of adding ammonia to this system. As you may recall, this was one of the experiments demonstrated on the first sequence of Videocassette 1. Adding ammonia solution and solid ammonium chloride* results in the formation of the complex $[Co(NH_3)_6]^{2+}(aq)$, which has a similar (red) colour to the aquo complex, $[Co(H_2O)_6]^{2+}$. To emphasise the change of ligand, the complexing process can be represented by the equation:

$$[Co(H_2O)_6]^{2+}(aq) + 6NH_3(aq) = [Co(NH_3)_6]^{2+}(aq) + 6H_2O(l) \quad \textbf{19}$$

With respect to the system represented by equation 18, the crucial point is that adding ammonia effectively removes $Co^{2+}(aq)$, by way of the process in equation 19.

□ What other species on the *right*-hand side of equation 18 is affected by adding ammonia?

■ $H^+(aq)$. Despite the presence of ammonium chloride (see the footnote), adding $NH_3(aq)$ inevitably lowers the concentration of $H^+(aq)$ to some extent, thereby raising the pH.

Now, *by themselves*, both these effects (amounting as they do to 'removal' of $H^+(aq)$ and $Co^{2+}(aq)$) would tend to move the equilibrium in equation 18 to the right, thus further stabilising cobalt in oxidation state +2. But in practice rather the reverse happens: provided a little activated charcoal is added to speed things along, bubbling air or oxygen through the ammoniacal solution changes its colour to yellow–brown. Again, you may recall that the colour is due to the ammine complex of cobalt(III), $[Co(NH_3)_6]^{3+}(aq)$: the complexing process can be represented by an equation directly analogous to that for $Co^{2+}(aq)$ (equation 19), as:

$$[Co(H_2O)_6]^{3+}(aq) + 6NH_3(aq) = [Co(NH_3)_6]^{3+}(aq) + 6H_2O(l) \quad \textbf{20}$$

To draw these observations together, suppose the cobalt(III)/cobalt(II) system in equation 18 is written in a more general form, as:

$$2Co^{III} + H_2O(l) = 2Co^{II} + 2H^+(aq) + \tfrac{1}{2}O_2(g) \quad \textbf{21}$$

Evidently this equilibrium lies to the *right* when *water* is the ligand: under these circumstances, cobalt(II) is stable with respect to cobalt(III) in an oxygenated acid solution. But if ammonia is added, $[Co(NH_3)_6]^{2+}(aq)$ is formed, and oxidised by oxygen to $[Co(NH_3)_6]^{3+}(aq)$. In other words, the equilibrium in equation 21 lies to the *left* when *ammonia* is the ligand. We conclude that changing the ligand from water to ammonia stabilises cobalt(III) with respect to cobalt(II).

* Aqueous ammonia is a weak base by virtue of the reaction:

$$NH_3(aq) + H_2O(l) = NH_4^+(aq) + OH^-(aq)$$

The added ammonium ions displace this equilibrium to the left, and thereby help both to maintain high concentrations of the complexing agent, $NH_3(aq)$, and to limit increase in the hydroxide ion concentration. The latter effect prevents possible formation of hydroxide precipitates (see SAQs 8 and 9, at the end of Section 4.2).

4.2 Quantifying the argument—stability constants

Our next step must be to place this conclusion—largely drawn from qualitative observations—on a proper quantitative footing. A clue to the approach we shall take is contained in the language used above, which is based (albeit implicitly) on an application of Le Chatelier's principle. Thus, we argued that complex formation by Co^{2+}(aq) with ammonia (equation 19) effectively removes Co^{2+}(aq) from the equilibrium system in equation 18. This, together with the lowering of the hydrogen ion concentration, would tend to stabilise cobalt(II). The final outcome, the stabilisation of cobalt(III), must mean that 'removal' of Co^{3+}(aq) by the analogous complexing reaction (equation 20) is sufficient to offset both of these effects, hence 'tipping the balance' in equation 18 over to the left.

SFC 1

These arguments are already couched in broadly thermodynamic terms: our task is to make this more precise.

□ What thermodynamic quantity characterises the equilibrium position in a reaction?

■ The *equilibrium constant*, hitherto usually denoted as K.

You may (with equal merit) have answered, 'the value of $\Delta G_m^\ominus$ for the reaction': the two quantities are, of course, related. In the Second Level Course, we wrote the formal relation between them as follows:

SLC 1

$$\Delta G_m^\ominus = -2.303RT \log K \qquad \mathbf{22}$$

where R ($=8.314\,J\,K^{-1}\,mol^{-1}$) is the gas constant.

For our present purposes, however, it is more revealing to concentrate on the equilibrium constant, rather than $\Delta G_m^\ominus$. Thus, a particularly fruitful step forward comes by recognising that the equilibrium constant for a complexing reaction, like equation 19 or equation 20, has a rather special significance: *it provides a measure of the stability of the complex in question, with respect to the ligand and the corresponding aqueous ion.* Not surprisingly, it is called the **stability constant** (or sometimes the **formation constant**) of the complex.*

By convention, the formation reaction is generally written without drawing any distinction between the six water molecules bound in the aqueous complex and the other water molecules in the bulk of the solution. Thus, equation 19, for example, becomes:

$$Co^{2+}(aq) + 6NH_3(aq) = [Co(NH_3)_6]^{2+}(aq) \qquad \mathbf{23}$$

Write an expression for the stability constant of the cobalt(II) complex, $[Co(NH_3)_6]^{2+}$(aq), in terms of the concentrations of the species in equation 23.

SFC 2

As you know, the form of an equilibrium constant follows directly from the stoichiometry of the balanced reaction equation *as written*. From equation 23 the stability constant of $[Co(NH_3)_6]^{2+}$ is given by:

$$K = \frac{c\{[Co(NH_3)_6]^{2+}\}}{c(Co^{2+})\{c(NH_3)\}^6} \qquad \mathbf{24}$$

where use of the symbol c(X) to denote the concentration of species X helps to avoid confusion with the use of square brackets to delineate a complex ion.

Now look through our discussion of the effect of adding ammonia to the system in equation 18:

$$2Co^{3+}(aq) + H_2O(l) = 2Co^{2+}(aq) + 2H^+(aq) + \tfrac{1}{2}O_2(g) \qquad \mathbf{18}$$

What do you conclude about the relative sizes of the stability constants for the two complex ions, $[Co(NH_3)_6]^{3+}$(aq) and $[Co(NH_3)_6]^{2+}$(aq)?

* In your reading, you may find the symbol β (Greek beta) used to represent the stability constant of a complex as defined here. We shall eschew such complications, and stick with the general symbol K.

The expression in equation 24 is typical of that for a stability constant. Evidently, the larger the stability constant for a given complex, the more effectively the ligand in question will 'mop up' the corresponding aqueous ion. Thus, the stability constant of the cobalt(III) complex must be greater than that of the cobalt(II) complex. In practice, it is greater by a factor of some 10^{31}: more than sufficient, it transpires, to offset also the effect of rising pH.

The final sentence is a reminder that there are two effects at work in the system considered here. It is important to keep these separate. Firstly, there is the effect of complex formation on the relative stability of the +2 and +3 oxidation states of cobalt; secondly, there is the effect of changing pH. The thrust of the discussion above is that the larger stability constant of $[Co(NH_3)_6]^{3+}(aq)$ stabilises cobalt in the +3 state. Now, if this is so, it should also be reflected in the electrode potential of the 'ammine' couple, $E^{\ominus}([Co(NH_3)_6]^{3+} | [Co(NH_3)_6]^{2+})$: after all, it is this quantity that provides a measure of the 'oxidising power' of a given redox couple.

□ Do the following values of $E^{\ominus}$ support the conclusion drawn above?

$$Co^{3+}(aq) + e = Co^{2+}(aq); \quad E^{\ominus} = 1.94\,V \qquad 16$$

$$[Co(NH_3)_6]^{3+}(aq) + e = [Co(NH_3)_6]^{2+}(aq); \quad E^{\ominus} = 0.1\,V \qquad 25$$

■ Yes. We have argued that the more positive the value of $E^{\ominus}(M^{3+} | M^{2+})$, the more stable is M^{2+} with respect to M^{3+}. Thus, the *lowering* of $E^{\ominus}$ (by 1.84 V) represents a substantial stabilisation of the +3 state.

Notice that we now appear to have *two* descriptions of the stabilisation of cobalt(III)—the one in terms of Le Chatelier's principle and the stability constants of the ammine complexes, and the other in terms of the lowering of the electrode potential. Since both descriptions invoke thermodynamic quantities, we might expect them to be connected in some way. This is indeed the case. As you will see in the next Section, the link is provided by the *Nernst equation.* For our purposes, the importance of this result lies in the ease with which it handles both the direct effects of complex formation (as outlined above), and other more *indirect* effects—notably changes in pH, the second factor in the system considered here.

Before turning to the next Section, make sure you work through the following SAQs. SAQs 7 and 8 revise and consolidate what you already know about equilibrium constants from previous courses. They also raise one or two problems that are taken up below. SAQ 9 extends the analysis in this Section to a system where the relative stability of oxidation states is affected, not by complexing, but by the formation of sparingly soluble compounds.

SAQ 7 (*revision*) Write equilibrium constant expressions for each of the following reactions, in terms of the concentrations of reactants and products:

(*a*) $Co^{3+}(aq) + 6NH_3(aq) = [Co(NH_3)_6]^{3+}(aq)$ **26**

(b) $2Co^{3+}(aq) + H_2(g) = 2Co^{2+}(aq) + 2H^+(aq)$ **27**

(c) $Co(OH)_2(s) = Co^{2+}(aq) + 2OH^-(aq)$ **28**

Indicate where you see a problem or ambiguity with the term 'concentration' as applied to any particular species in the reaction.

SAQ 8 (*revision*) As you may recall from Block 1, a blue precipitate of cobalt(II) hydroxide, $Co(OH)_2$, forms when hydroxide ions are added to a solution containing $Co^{2+}(aq)$. The **solubility product**, K_{sp}, of a sparingly soluble salt like this was discussed in a Second Level Course. Using the definition introduced there, K_{sp} for $Co(OH)_2$ is written:

SLC 3

$$K_{sp} = \{c(Co^{2+})\}\{c(OH^-)\}^2 \qquad 29$$

Use thermodynamic data from the S343 *Data Book* to calculate $\Delta G_m^{\ominus}$ at 298.15 K for reaction 28 in part (c) of SAQ 7, and hence the corresponding value of the equilibrium constant (using equation 22). Do you see a problem in identifying your calculated value with K_{sp} for $Co(OH)_2$?

SAQ 9 You may further recall that the initial precipitate of $Co(OH)_2$ referred to in SAQ 8 gradually darkens in air. This is attributed to oxidation by atmospheric oxygen to the corresponding cobalt(III) compound, which can be written $Co(OH)_3$. Use the information in Table 5 to interpret these observations in terms of the equilibrium system discussed in Section 4:

$$2Co^{3+}(aq) + H_2O(l) = 2Co^{2+}(aq) + 2H^+(aq) + \tfrac{1}{2}O_2(g) \qquad \textbf{18}$$

Table 5 Properties of hydroxides of cobalt

Compound	Colour	K_{sp}
$Co(OH)_2$	pink or blue*	$10^{-15}\,mol^3\,l^{-3}$
$Co(OH)_3$	black	$10^{-45}\,mol^4\,l^{-4}$

* The colour depends on the conditions, but only the pink form is permanently stable.

5 THE NERNST EQUATION AND ITS APPLICATIONS

5.1 Introduction

In discussing the cobalt(III)/cobalt(II) system in the previous Section, we focused on the way that complex formation (or, as you found in answering SAQ 9, the formation of sparingly soluble compounds) affects the relative stability of the two oxidation states. Seen in a broader context, however, the example considered there is just one instance of the way changes in the *composition* of a system can affect the *direction* of chemical change. This in turn, suggests that we have entered the realm of the *second law of thermodynamics*—but an aspect of it that we have largely ignored up till now.

SLC 4

To make this more explicit, let us just recap on your first encounter with the second law. In a Second Level Course, we applied this universal law to a typical chemical system—a reaction taking place under the normal laboratory conditions of constant temperature and pressure. This analysis resulted in a simple criterion for a spontaneous reaction; we wrote it in terms of the Gibbs function as follows:

$$\Delta G < 0 \quad \text{(at constant } T \text{ and } p) \qquad \textbf{30}$$

SLC 5

We went on to assert that $\Delta G_m^\ominus$ is the molar value of ΔG when reactants and products are in their *standard states*, and thereafter adopted the criterion $\Delta G_m^\ominus < 0$ as our definition of a thermodynamically favourable reaction. This is the criterion we have continued to use throughout the present Course, although we have often chosen to express it in terms of standard electrode potentials, $E^\ominus$, by virtue of the relation:

$$\Delta G_m^\ominus = -nFE^\ominus \qquad \textbf{31}$$

The crucial point is that, strictly speaking, values of $E^\ominus$ (like the equivalent values of $\Delta G_m^\ominus$) afford a comparison of the oxidising power of redox couples when, *and only when*, all the species in the redox equation are in their standard states. Thus, for example, for the redox system

$$Co^{3+}(aq) + e = Co^{2+}(aq); \quad E^\ominus = 1.94\,V \qquad \textbf{16}$$

the value of 1.94 V is a thermodynamic measure of the 'strength' of $Co^{3+}(aq)$ as an oxidising agent when both $Co^{2+}(aq)$ and $Co^{3+}(aq)$ are in their standard states. To pre-empt our discussion a little, the standard state of a substance can be equated with its being present at *unit activity*; for an ionic solute in aqueous solution, like $Co^{2+}(aq)$ or $Co^{3+}(aq)$ for example, this can be taken as very roughly equivalent to a concentration of $1\,mol\,l^{-1}$.

A complexing ligand, say NH_3, if added to a solution of $Co^{2+}(aq)$ and $Co^{3+}(aq)$ at unit activity, mops up both ions and drastically lowers their concentrations. From this point of view, then, the effect of complex formation is to 'drive' the system in equation 16 away from its standard state. Under these circumstances, the quantity that provides a measure of the oxidising power of the couple is not $E^\ominus$, but E—where dropping the superscript indicates that we are no longer working at unit activity. As you will see shortly, the link between E and $E^\ominus$ is provided by the Nernst equation.

5.2 The Nernst equation: standard states and activities

For the purposes of this Course, it is important only that you recognise the Nernst equation—and know how to use it to analyse and discuss the sort of chemical problem outlined in Section 4. To this end, we shall not attempt to derive the equation here, nor explore its thermodynamic 'roots', as it were: rather we simply quote the final result.*

For a general redox couple, which can be written in 'alphabetical' form as follows:

$$aA + bB + \cdots + ne = xX + yY + \cdots \qquad 32$$

where n is the coefficient of e, the **Nernst equation** takes the form:

$$E = E^\ominus - \left(\frac{2.303RT}{nF}\right) \log \left\{\frac{a_X{}^x a_Y{}^y \cdots}{a_A{}^a a_B{}^b \cdots}\right\} \qquad 33$$

Here, the quantities (a_A, a_B ..., and a_X, a_Y, ..., etc.) in the logarithmic term are the **activities** of the reactants and products, respectively, each raised to the power of the appropriate stoichiometric number. Evidently, this expression does indeed provide a link between E and $E^\ominus$. But, to use this link in order to 'adjust' values of $E^\ominus$ to non-standard conditions, we must be able to determine the effect of the second term on the right-hand side in equation 33. This, in turn, raises the more fundamental question of what is meant by the 'activity' of a substance.

As suggested earlier, the two concepts—'activity' and 'standard state'—are intimately interrelated. In particular, the **standard state** of a substance can be equated with its being at **unit activity**.

□ Does the form of equation 33 accord with this idea?

■ Yes. If the activity of each substance in the redox couple (equation 32) is unity, then so too will be the values of $a_A{}^a$, ... $a_X{}^x$, ..., etc., and hence the ratio in the logarithmic term. Since log 1 = 0, the 'correction term' vanishes, $E = E^\ominus$ and the electrode potential takes its standard value.

In practice, the standard state is chosen (quite arbitrarily) to be the most convenient for the job in hand, and such that the activity of the substance in question can be readily related to measurable properties of that substance. It turns out (for reasons we cannot go into here) that this leads to definitions of activity that differ according to the *physical* state of the substance. To 'flesh out' these rather abstract ideas, we ask you to accept the following **definitions of activity** in three very important cases (all at a given temperature T).

Case 1 For a pure liquid or solid, $a = 1$ **34**

In other words, the activity of a pure liquid or solid is always unity: it is always in its standard state.

* If you are interested, a fairly full development is given in another third-level Course.

Case 2 For a pure gas (or gaseous component in a mixture), $a = p/p^{\ominus}$ **35**

where p is the pressure of the gas, and $p^{\ominus}$ is a standard pressure. Until very recently, $p^{\ominus}$ was universally chosen to be 1 atm. However, the recommendation of the International Union of Pure and Applied Chemistry (IUPAC) is that this be changed slightly—to 10^5 Pa or 1 bar (recall 1 atm = 101 325 Pa = $1.013\,25 \times 10^5$ Pa = 1.013 25 bar): accordingly, we have adopted this revised standard in this Course (see also the relevant Section of the S343 *Data Book*).

□ What does the definition in equation 35 imply about the standard state of a gas?

■ The standard state is established at unit activity. Here, $a = 1$ when $p = p^{\ominus}$, so a gas is in its standard state at a pressure of 1 bar.

Case 3 For a solute in aqueous solution, $a = c/c^{\ominus}$ **36**

where c is the molar concentration of the solute, and $c^{\ominus}$ is a standard concentration. Throughout this Course, we shall adopt the standard $c^{\ominus} = 1\,\text{mol}\,\text{l}^{-1}$, such that the standard state is established at a concentration $c = c^{\ominus} = 1\,\text{mol}\,\text{l}^{-1}$.

Notice that these definitions highlight a further aspect of the activity concept: it is a *dimensionless quantity*—it has no unit. Once again, the reasons for this lie deep within the formalism of thermodynamics, and we cannot go into them here. However, some insight can be gained by thinking of the activity as a measure of the 'departure' from the standard state—as suggested by the ratios in equations 35 and 36. Thus, for an aqueous solute, for example, the more its concentration 'departs' from $1\,\text{mol}\,\text{l}^{-1}$, the more different from unity is its activity.

Of the three definitions above, Case 1 is quite general: indeed we shall *assume* that it applies also to the *solvent* in a solution. On the other hand, the definitions in Cases 2 and 3 are more restricted: strictly speaking they apply *only to ideal gases and ideal solutions, respectively*.

Put very simply, in an ideal system (whether a gas or a solution) the individual particles (be they atoms, molecules or ions) behave independently of one another. At pressures little different from ambient, many real gases do indeed show this type of behaviour—to a good approximation at least. But for real solutions—and in particular, those containing ionic species—the situation is very different. Here, the assumption of ideal behaviour is often a poor approximation, unless, that is, the solution is very dilute indeed (usually less than $10^{-3}\,\text{mol}\,\text{l}^{-1}$ for electrolyte solutions). Nevertheless, we shall use the definition of activity in equation 36 throughout this Course, no matter how concentrated the solution in question. Our chief justification is that neglecting so-called 'deviations from ideality' rarely leads to predictions about *chemical* behaviour that are at odds with experiment—at least in so far as the relative stability of oxidation states is concerned.

To gain some familiarity with the Nernst equation, try the following SAQ.

SAQ 10 Use the Nernst equation and the definitions in equations 34 to 36, as appropriate, to write expressions for the electrode potentials of the following couples, in terms of the concentrations and/or pressures of the species involved:

(a) $Co^{3+}(aq) + e = Co^{2+}(aq)$ **16**

(b) $\frac{1}{2}O_2(g) + 2H^+(aq) + 2e = H_2O(l)$ **17**

Given $R = 8.314\,\text{J}\,\text{K}^{-1}\,\text{mol}^{-1}$ and $F = 96\,485\,\text{C}\,\text{mol}^{-1}$, what is the value of the constant term in the Nernst equation at 298.15 K? (Recall the definition of the volt, $\text{V} = \text{kg}\,\text{m}^2\,\text{s}^{-3}\,\text{A}^{-1} = \text{J}\,\text{A}^{-1}\,\text{s}^{-1} = \text{J}\,\text{C}^{-1}$.)

To summarise: at 298.15 K, the Nernst equation reduces to the following simplified form:

$$E = E^{\ominus} - \left(\frac{0.059\,2\,\text{V}}{n}\right) \log \left\{\frac{a_X{}^x\, a_Y{}^y \cdots}{a_A{}^a\, a_B{}^b \cdots}\right\} \qquad \textbf{37}$$

The activities $a_A, a_B, \ldots, a_X, a_Y, \ldots$, etc., are dimensionless quantities: for pure solids and liquids, they are unity; for gases, they are roughly equal to the numerical value of the pressure in bars (and roughly equal to that in atmospheres, since $1 \text{ atm} \approx 1 \text{ bar}$); for aqueous solutes, they are roughly equal to the numerical value of the concentration in mol l^{-1}.

There remains just one loose end to be dealt with before we can apply the Nernst equation to the problem discussed in Section 4.

5.3 Equilibrium constants revisited—an aside

The answers to SAQs 7 and 8 (Section 4.2) highlighted one or two problems about the dimensions of equilibrium constants calculated from thermodynamic data—as opposed to those determined experimentally. The definitions of activity introduced above allow us to resolve these difficulties very simply. First, we note that the dimensionless equilibrium constant defined by equation 22,

$$\Delta G_m^\ominus = -2.303RT \log K \qquad 22$$

is formally called the **standard** or **thermodynamic equilibrium constant**, and represented by the symbol $K^\ominus$. Thus, equation 22 should strictly be written:

$$\Delta G_m^\ominus = -2.303RT \log K^\ominus \qquad 38$$

It follows that equilibrium constants calculated from thermodynamic data are actually values of $K^\ominus$ (at the temperature in question). To take the example in SAQ 8, for instance:

$$Co(OH)_2(s) = Co^{2+}(aq) + 2OH^-(aq) \qquad 28$$

$$K^\ominus(298.15\,\text{K}) = 1.1 \times 10^{-15}$$

The link between $K^\ominus$ and the more familiar 'experimental' equilibrium constants—usually expressed in terms of concentrations, and hence with dimensions dependent on the stoichiometry of the reaction equation—is provided by the formal definition of $K^\ominus$. Thus, $K^\ominus$ has the normal form of an equilibrium constant (to paraphrase: 'products over reactants'), but expressed as a *ratio of activities*: since the latter are dimensionless, so too is $K^\ominus$.

□ Write an expression for $K^\ominus$ for the reaction in equation 28.

■ $$K^\ominus = \frac{a(Co^{2+})\{a(OH^-)\}^2}{a\{Co(OH)_2\}} \qquad 39$$

Since the activity is unity for the solid, $Co(OH)_2$, the problem of 'concentration' as applied to a solid phase (SAQ 7) disappears. Using the definition of activity in equation 36, $K^\ominus$ then becomes

$$K^\ominus = \{c(Co^{2+})/c^\ominus\}\{c(OH^-)/c^\ominus\}^2 \qquad 40$$

which can be compared with the 'conventional' solubility product,

$$K_{sp} = \{c(Co^{2+})\}\{c(OH^-)\}^2 \qquad 29$$

Providing concentrations are expressed in the unit mol l^{-1}, it follows that the value of $K^\ominus$ cited above (1.1×10^{-15}) is just the numerical value of K_{sp} (at 298.15 K, of course); that is,

$$K_{sp} = 1.1 \times 10^{-15}\,\text{mol}^3\,\text{l}^{-3}$$

□ What asumption is involved in identifying the values of $K^\ominus$ and K_{sp}?

■ The ideal behaviour assumption implicit in using the definition of activity in Case 3.

This assumption underlies much of the discussion here and in later Sections: to avoid becoming tedious, we shall not constantly repeat it! However, you should bear in mind that it renders many of our statements only approximately true.

The arguments rehearsed above are not, of course, restricted to solubility products. Rather, *values of $K^{\ominus}$ derived from thermodynamic data can always be identified with the numerical value of the corresponding experimental equilibrium constant.** And the converse also holds. A case in point is the stability constant of the cobalt(III)–ammine complex, $[Co(NH_3)_6]^{3+}(aq)$: the experimental value is known to be around $10^{35}\,mol^{-6}\,l^6$ (at 298.15 K).

☐ Write an expression for the corresponding standard equilibrium constant.

■ The stability constant refers to the reaction in equation 26 (SAQ 7(a)):

$$Co^{3+}(aq) + 6NH_3(aq) = [Co(NH_3)_6]^{3+}(aq) \tag{26}$$

$$\text{So } K^{\ominus} = \frac{a([Co(NH_3)_6]^{3+})}{a(Co^{3+})\{a(NH_3)\}^6} \tag{41}$$

$$= \frac{\{c([Co(NH_3)_6]^{3+})/c^{\ominus}\}}{\{c(Co^{3+})/c^{\ominus}\}\{c(NH_3)/c^{\ominus}\}^6}$$

As before, comparison with the expression for the experimental stability constant (see the answer to SAQ 7(a)) confirms that if $c^{\ominus} = 1\,mol\,l^{-1}$ we may write $K^{\ominus} = 10^{35}$. From now on, we shall often assume that $K^{\ominus}$ and K (or K_{sp}) are equivalent, simply adding or striking out the units as appropriate. Equally, we may sometimes be a touch imprecise about the term 'activity', using it and 'concentration' fairly interchangeably.

SAQ 11 Given the quoted 'experimental' equilibrium constants for the following reactions (both at 298.15 K), what is the corresponding value of $K^{\ominus}$ in each case?

(a) $Co(OH)_3(s) = Co^{3+}(aq) + 3OH^-(aq)$; $K_{sp} = 10^{-45}\,mol^4\,l^{-4}$;

(b) $N_2O_4(g) = 2NO_2(g)$; $K = 1.5 \times 10^4\,Pa$.

With this brief digression over, we are now in a position to return to the Nernst equation and its manifold applications.

5.4 Using the Nernst equation

5.4.1 Electrode potentials and complex formation

The first example to hand is the cobalt(III)/cobalt(II) system: recall that we closed the discussion in Section 4.2 by quoting $E^{\ominus}$ values for both the aquo and the ammine couples, as:

$$Co^{3+}(aq) + e = Co^{2+}(aq); \quad E^{\ominus} = 1.94\,V \tag{16}$$

$$[Co(NH_3)_6]^{3+}(aq) + e = [Co(NH_3)_6]^{2+}(aq); \quad E^{\ominus} = 0.1\,V \tag{25}$$

Now, in line with the discussion in Section 5.2, 0.1 V can obviously be regarded as the standard potential of the ammine couple, appropriate to a solution containing both complex ions, each at unit activity. But such a solution must also contain finite (albeit very small) concentrations of the two aquo ions, $Co^{2+}(aq)$ and $Co^{3+}(aq)$, by virtue of the equilibria described by the stability constants of the two complexes. This, in turn, suggests an alternative interpretation of the value 0.1 V, namely that it represents the potential of the $Co^{3+}(aq)/Co^{2+}(aq)$ couple, but under conditions markedly different from those to which the standard value (1.94 V) applies. As we shall see, the two values are, in fact, related via the Nernst equation, and the stability constants of the ammine complexes.

* ... with the additional proviso that pressures of gases are expressed in bars, so that a pressure is, say, x bar, and $p/p^{\ominus} = x$. As mentioned earlier, expressing pressures in atmospheres produces roughly the same result because 1 bar and 1 atm are very similar. See SAQ 11 below for an example.

According to the answer to SAQ 10(a), the Nernst equation for the aquo couple (equation 16) becomes:

$$E = E^{\ominus} - (0.059\,2\text{ V})\log\{a(\text{Co}^{2+})/a(\text{Co}^{3+})\}$$

$$= 1.94\text{ V} - (0.059\,2\text{ V})\log\{c(\text{Co}^{2+})/c(\text{Co}^{3+})\} \qquad \textbf{42}$$

□ Now use information from Section 5.3 to write an expression for the concentration of Co^{3+}(aq) in a solution containing the complex $[Co(NH_3)_6]^{3+}$(aq) at unit activity.

■ Using the symbol $K^{\ominus}_{+3}$ to indicate the stability constant of $[Co(NH_3)_6]^{3+}$(aq):

$$K^{\ominus}_{+3} = \frac{a([\text{Co(NH}_3)_6]^{3+})}{a(\text{Co}^{3+})\{a(\text{NH}_3)\}^6}$$

With $a([Co(NH_3)_6]^{3+}) = 1$, rearrangement gives:

$$a(\text{Co}^{3+}) = \frac{c(\text{Co}^{3+})}{c^{\ominus}} = \frac{1}{K^{\ominus}_{+3}\{a(\text{NH}_3)\}^6}$$

$$\text{or } c(\text{Co}^{3+}) = \frac{c^{\ominus}}{K^{\ominus}_{+3}\{a(\text{NH}_3)\}^6} \qquad \textbf{43}$$

Likewise, the stability constant of the cobalt(II) complex can be written $K^{\ominus}_{+2}$, such that

$$c(\text{Co}^{2+}) = \frac{c^{\ominus}}{K^{\ominus}_{+2}\{a(\text{NH}_3)\}^6} \qquad \textbf{44}$$

Combining these results and substituting into equation 42, gives

$$E = 1.94\text{ V} - (0.059\,2\text{ V})\log(K^{\ominus}_{+3}/K^{\ominus}_{+2}) \qquad \textbf{45}$$

Recalling that $K^{\ominus}_{+3} = 10^{35}$ and noting that $K^{\ominus}_{+2}$ is around 10^4, this becomes

$$E = 1.94\text{ V} - (0.059\,2\text{ V})\log(10^{35}/10^4) \qquad \textbf{46}$$

$$= 1.94\text{ V} - 1.84\text{ V}$$

$$= 0.10\text{ V}$$

in accord with the value of $E^{\ominus}$ for the ammine couple quoted earlier.

There are several points to note here. First, the analysis outlined above is based on completely general principles: in particular, it can be applied whenever a metal forms *both* aquo ions in two different oxidation states, *and* corresponding complex ions with a given ligand. Provided the stability constants of the complex ions are known, the standard potential of the 'complex' couple—call it $E^{\ominus}_{comp}$, say—can always be calculated from that of the aquo couple, as above. (A more general expression for $E^{\ominus}_{comp}$ is given in Section 5.5.) Because of this link, $E^{\ominus}$ of the aquo couple provides a kind of base line against which to measure the ability of a given ligand to stabilise one oxidation state with respect to the other. Thus, the calculation embodied in equation 42 or equation 45 is a concrete expression of the dramatic stabilisation of the +3 state of cobalt by ammonia.

This highlights a second, related point: the stability constants of complexes may differ by many orders of magnitude—a factor of 10^{31} in our example. Thus, despite the logarithmic dependence in the Nernst equation, the stability of a given oxidation state may, as you have seen, change very markedly on the addition of a ligand that forms stable complexes, or, as you should prove for yourself by working through the following SAQ, precipitates sparingly soluble compounds.

SAQ 12 A solution contains unit activity (1 mol l^{-1}) of Co^{2+}(aq) and Co^{3+}(aq), when $E(Co^{3+}|Co^{2+}) = E^{\ominus}(Co^{3+}|Co^{2+}) = 1.94$ V. Suppose now that hydroxide ions are added to precipitate $Co(OH)_2$(s) and $Co(OH)_3$(s), until the pH of the solution steadies at 14 (that is, $a(OH^-) = 1$). Given that the standard solubility products of $Co(OH)_2$ and $Co(OH)_3$ are around 10^{-15} and 10^{-45}, respectively, what are the concentrations of Co^{2+}(aq) and

$Co^{3+}(aq)$ in the resulting solution? Hence calculate the *standard* potential, $E^{\ominus}$, of the following couple:

$$Co(OH)_3(s) + e = Co(OH)_2(s) + OH^-(aq) \qquad \textbf{47}$$

5.4.2 The pH-dependence of electrode potentials—'solvent decomposition'

As suggested by the example in Section 4, another important factor in any aqueous solution is the concentration of hydrogen ions. Since this may vary by a factor of around 10^{14} in water, a change in pH may cause a marked alteration in the behaviour of particular oxidation states, *even where it does not change the nature of the complexes involved.*

In this context, a particularly important example is the other couple that played a key role in Section 4—the 'so-called' *oxygen electrode*:

$$\tfrac{1}{2}O_2(g) + 2H^+(aq) + 2e = H_2O(l); \quad E^{\ominus} = 1.23\,V \qquad \textbf{17}$$

As you should have found in answering SAQ 10(b), the Nernst equation for this couple can be written

$$E = E^{\ominus} - \left(\frac{0.059\,2\,V}{2}\right)\log\left(\frac{1}{\{p(O_2)/p^{\ominus}\}^{1/2}\{c(H^+)/c^{\ominus}\}^2}\right)$$

If we assume (as we shall throughout this Block) that the pressure of oxygen is little different from ambient, such that $p(O_2) \approx p^{\ominus} \approx 1$ atm and $\{p(O_2)/p^{\ominus}\}^{1/2} = 1$, this becomes*

$$E = 1.23\,V + (0.059\,2\,V)\log\{c(H^+)/c^{\ominus}\} \qquad \textbf{48}$$

Introducing the following more precise definition of pH:

$$pH = -\log a(H^+) \quad \textit{(by definition)} \qquad \textbf{49}$$
$$= -\log\{c(H^+)/c^{\ominus}\}$$

equation 48 then reduces to the following simple form:

$$E/V = 1.23 - 0.059\,2\,pH \qquad \textbf{'oxygen electrode'} \qquad \textbf{50}$$

As noted earlier in the Course, the importance of this potential is twofold: *it defines both the resistance of water to oxidation, and the 'oxidising power' of atmospheric oxygen in contact with an aqueous solution.* At that stage, our discussion was restricted to acid solutions (strictly at zero pH—unit activity of hydrogen ions), to which the standard value applies. Under these circumstances, we argued that the oxidised states of all couples with potentials greater than 1.23 V should oxidise water (the obvious example being the instability of $Co^{3+}(aq)$), whereas the reduced states of all couples with potentials below this value should be oxidised by dissolved oxygen (examples being the air-sensitive ions, $Cr^{2+}(aq)$ and $V^{2+}(aq)$, with $E^{\ominus}(M^{3+}(aq)\,|\,M^{2+}(aq))$ values of -0.42 V and -0.26 V, respectively).

□ What other potential is also important in determining the stability of lower oxidation states—such as $Cr^{2+}(aq)$ and $V^{2+}(aq)$ for example—in aqueous solution?

■ The potential of the hydrogen electrode:

$$H^+(aq) + e = \tfrac{1}{2}H_2(g); \quad E^{\ominus} = 0\,V \qquad \textbf{51}$$

which defines the resistance to reduction of the hydrogen ions in an aqueous solution.

* There are two points to note; firstly, you may argue—and quite rightly—that the ambient pressure of O_2 is not 1 atm, but around 0.2 atm. However, this introduces a correction (around 0.01 V), which is insignificant compared with the effect of changing pH. Secondly, in obtaining equation 48, we have used the following property of logarithms:

$$\log\frac{1}{x^n} = \log x^{-n} = -n\log x$$

where in this case $n = 2$.

Assuming unit activity of hydrogen gas, the Nernst equation leads to the following expression for the pH-dependence of this potential:

$$E/\mathrm{V} = -0.059\,2\,\mathrm{pH} \qquad \textbf{`hydrogen electrode'} \qquad 52$$

Thus, the Nernst equation allows us to extend the analysis of aqueous redox chemistry to neutral and alkaline solutions. In this context, the expressions in equations 50 and 52 can be seen to effectively limit the *thermodynamic* stability of oxidising or reducing agents in aqueous solution to couples whose potentials lie in the range $(-0.059\,2\,\mathrm{pH})$ V to $(1.23 - 0.059\,2\,\mathrm{pH})$ V.

What actually happens in practice, of course, often depends just as much on *kinetic factors*. Recall one of the examples cited in Block 1: in most laboratory experiments, acidified solutions of Fe^{2+}(aq) can be handled in air without serious deterioration, *despite* the fact that $E^{\ominus}(Fe^{3+}(aq)|Fe^{2+}(aq)) = 0.77$ V, such that Fe^{2+}(aq) is thermodynamically unstable to aerial oxidation under these conditions.

SAQ 13 In SAQ 12, you calculated the standard potential of the following couple:

$$Co(OH)_3(s) + e = Co(OH)_2(s) + OH^-(aq); \quad E^{\ominus} = 0.16\,\mathrm{V} \qquad 47$$

(a) Calculate E for the oxygen electrode at pH 10.

(b) Calculate E for the cobalt hydroxides electrode (equation 47) at pH 10.

Refer now to the experiment described in SAQs 8 and 9 at the end of Section 4.2. Suppose that adding hydroxide to a solution of Co^{2+}(aq) results in a pink precipitate of $Co(OH)_2$ in contact with a solution of pH 10. Would you expect the precipitate to darken due to aerial oxidation to $Co(OH)_3$?

5.4.3 The relative stability of oxidation states—drawing the threads together

According to the expressions in equations 50 and 52, the potentials of the oxygen and hydrogen couples fall (become more negative) as the pH is raised. This, in turn, suggests that low oxidation states ('reduced' states like Fe^{2+}(aq) or Co^{2+}(aq), for example) may be stabilised relative to higher ones in alkaline solutions. But this is certainly not the case for cobalt (SAQs 12 and 13); nor is it true for other transition metals, such as chromium and manganese, for example. Indeed, you may recall (from Block 1) that the stabilisation of higher oxidation states in alkaline media is a characteristic feature of the chemistry of these elements!

In some cases, like cobalt, raising the pH precipitates hydroxides (or oxides), the solubilities of which generally drop off very markedly with increasing oxidation state of the metal. By contrast, high oxidation states of metals like chromium and manganese commonly exist as oxoanions in solution, one example being the permanganate ion, MnO_4^-(aq), containing manganese(VII). Here, the electrode potential for reduction to a lower oxidation state is nearly always pH-dependent, sometimes markedly so, as for example:

$$MnO_4^-(aq) + 8H^+(aq) + 5e = Mn^{2+}(aq) + 4H_2O(l) \qquad 53$$

In either of these situations, the potential of the couple containing two oxidation states of the metal—Co^{III}/Co^{II}, for example, or Mn^{VII}/Mn^{II} in equation 53—is itself lowered by raising the pH.* Furthermore, and crucially, this change in potential is often *greater* than that in the corresponding potentials appropriate to the oxidation or reduction of water. The net effect is to *destabilise* the lower oxidation state in alkaline media—leading to the observed stabilisation of higher states, as in the Co^{III}/Co^{II} system. (There are further examples for you to work through in SAQs 15 and 16.)

* If you look through the examples above, you will see that this follows *whenever* the electrode reaction has H^+(aq) on the left or (equivalently) OH^-(aq) on the right.

Finally, we close this Section by returning to the Co^{III}/Co^{II}–ammine system: by this stage, it should be clear that the Nernst equation does indeed allow us to quantify both the direct and indirect effects described in Section 4. Thus, the standard potential of the aquo couple (1.94 V) is well above the value of 1.23 V required for the oxidation of water in acid solution. At room temperature, the reaction is also kinetically favourable, so Co^{3+}(aq) causes rapid decomposition of the solvent.

Adding ammonia produces two effects. Firstly, the *nature* of the cobalt species changes: ammine complexes are formed, and the potential of the Co^{III}/Co^{II} couple plummets to 0.1 V (Section 5.4.1). Secondly, the pH of the solution rises: a typical value in the final solution being about 8. Under these circumstances, the potential of the oxygen electrode is $E = (1.23 - 0.059\,2 \times 8)\,\text{V} = 0.76\,\text{V}$. Since this is larger (more positive) than the potential of the cobalt–ammine couple, oxygen in contact with an aqueous solution of pH 8 is thermodynamically capable of oxidising $[Co(NH_3)_6]^{2+}$(aq) to the tripositive state. And so it does!

5.5 Summary of Section 5

1 The Nernst equation allows literature values of $E^\ominus$ to be adjusted to non-standard conditions. The most important points are collected in the following box.

Nernst equation

For the following general redox couple (at 298.15 K)

$$a\text{A} + b\text{B} + \cdots + ne = x\text{X} + y\text{Y} + \cdots \quad (32)$$

$$E = E^\ominus - \left(\frac{0.059\,2\,\text{V}}{n}\right) \log \left\{\frac{a_\text{X}{}^x\, a_\text{Y}{}^y \cdots}{a_\text{A}{}^a\, a_\text{B}{}^b \cdots}\right\} \quad (37)$$

where n is the number of electrons transferred, and the activities a are dimensionless quantities, defined according to the physical state of the species involved. To summarise:

(a) For a liquid or solid: $a = 1$.

(b) For a gas: $a = p/p^\ominus$ ($p^\ominus = 10^5\,\text{Pa} = 1\,\text{bar}$), that is, roughly the value of the pressure in atmospheres (since $1\,\text{atm} \approx 1\,\text{bar}$).

(c) For a solute in aqueous solution: $a \approx c/c^\ominus$ ($c^\ominus = 1\,\text{mol}\,\text{l}^{-1}$), that is, roughly the value of the concentration in $\text{mol}\,\text{l}^{-1}$.

2 Suppose that a metal forms aqueous ions in two different oxidation states, M^{2+}(aq) and M^{3+}(aq) being a familiar example. Suppose further that a ligand (L, say) forms new complexes with *both* aqueous ions, $[ML_6]^{2+}$(aq) and $[ML_6]^{3+}$(aq), for example. Then the relative thermodynamic stability of the complex ions is measured by $E^\ominus_\text{comp}$, the standard potential of the couple

$$[ML_6]^{3+}(\text{aq}) + e = [ML_6]^{2+}(\text{aq}) \quad (54)$$

3 The Nernst equation provides a link between $E^\ominus_\text{comp}$ and $E^\ominus$ for the corresponding aquo ions. For the system in equation 54:

$$E^\ominus_\text{comp} = E^\ominus(M^{3+}\,|\,M^{2+}) - (0.059\,2\,\text{V}) \log \{c(M^{2+})/c(M^{3+})\} \quad (55)$$

where $c(M^{2+})$ and $c(M^{3+})$ are the concentrations (strictly activities) of M^{2+}(aq) and M^{3+}(aq) ions, respectively, in a solution containing unit activities of the corresponding complex ions, $[ML_6]^{2+}$(aq) and $[ML_6]^{3+}$(aq). These concentrations can be evaluated from the stability constants of the complex ions, $K^\ominus_{+2}$ and $K^\ominus_{+3}$, respectively. To emphasise this connection, equation 55 can be rewritten:

$$E^\ominus_\text{comp} = E^\ominus(M^{3+}\,|\,M^{2+}) - (0.059\,2\,\text{V}) \log (K^\ominus_{+3}/K^\ominus_{+2}) \quad (56)$$

4 By virtue of the relation in equation 55 or 56, the value of $E^\ominus$ for the aquo couple provides a base line against which to measure the ability of a given ligand to stabilise one oxidation state with respect to the other. In brief:

> If $E^\ominus_{comp}$ is smaller (more negative, or less positive) than $E^\ominus(M^{3+} \mid M^{2+})$, then $K^\ominus_{+3} > K^\ominus_{+2}$ (that is, the ligand stabilises the higher oxidation state), and vice versa.

5 A similar analysis holds when the added ligand results in the precipitation of insoluble compounds, the only difference being that the concentrations to be inserted in equation 55 are now determined by the solubility products of the compounds in question.

6 The potentials of the oxygen and hydrogen couples:

$$\tfrac{1}{2}O_2(g) + 2H^+(aq) + 2e = H_2O(l) \qquad \textbf{17}$$

$$H^+(aq) + e = \tfrac{1}{2}H_2(g) \qquad \textbf{51}$$

namely $(1.23 - 0.0592\,pH)$ V and $(-0.0592\,pH)$ V, respectively, set limits on the thermodynamic stability of oxidising or reducing agents in aqueous solutions of given pH.

7 The type of analysis indicated in points 5 and 6 permits a rationalisation of the observation that high oxidation states are frequently stabilised in alkaline media (see also SAQs 14–16).

To enlarge and consolidate your understanding of the points summarised above, make sure you try the following SAQs at some point.

SAQ 14 Examine the following redox potentials:

$$Fe^{3+}(aq) + e = Fe^{2+}(aq); \quad E^\ominus = 0.77\,V \qquad \textbf{57}$$

$$Fe(OH)_3(s) + e = Fe(OH)_2(s) + OH^-(aq); \quad E^\ominus = -0.58\,V \qquad \textbf{58}$$

$$\tfrac{1}{2}O_2(g) + 2H^+(aq) + 2e = H_2O(l); \quad E = (1.23 - 0.0592\,pH)\,V \qquad \textbf{59}$$

(a) Use the Nernst equation to determine the ratio $K^\ominus_{sp}\{Fe(OH)_3\}/K^\ominus_{sp}\{Fe(OH)_2\}$ of the solubility products of iron(III) and iron(II) hydroxides, and hence (using information from Table 6) the value of $K^\ominus_{sp}\{Fe(OH)_3\}$.

(b) When an acidified solution of $Fe^{2+}(aq)$ is made alkaline, the initial precipitate is pale green, but quickly darkens on standing in air. Is this observation consistent with the information collected above? In what way does your explanation differ from that advanced to interpret the analogous experiment with $Co^{2+}(aq)$; see SAQs 8, 9, 12 and 13?

Table 6 Properties of hydroxides of iron

Compound	Colour	$K^\ominus_{sp}$
$Fe(OH)_2$	pale green*	10^{-16}
$Fe(OH)_3$	brown	

* Colourless when completely pure, but green as normally precipitated.

SAQ 15 In Block 1 we quoted the following process as the major reaction leading to the formation of manganese nodules:

$$Mn^{2+}(aq) + \tfrac{1}{2}O_2(g) + H_2O(l) = MnO_2(s) + 2H^+(aq) \qquad \textbf{60}$$

We claimed that the equilibrium lies to the left in acid, but that it lies well to the right at the pH of sea-water, that is, at a pH of around 8. Given that a typical concentration of $Mn^{2+}(aq)$ in sea-water is about $10^{-8}\,mol\,l^{-1}$, use the following electrode potential (together with any other relevant information) to justify this contention:

$$MnO_2(s) + 4H^+(aq) + 2e = Mn^{2+}(aq) + 2H_2O(l); \quad E^\ominus = 1.23\,V \qquad \textbf{61}$$

SAQ 16 The potential diagram for manganese in acid solution that you met in Block 1 is repeated in Figure 8a. At this stage, you should recognise that many of the couples in this diagram have potentials that are dependent on pH: the values quoted for acid solutions are calculated at pH 0 (unit activity of hydrogen ions). As you might expect, an entirely equivalent diagram can be drawn up for alkaline solutions, by calculating corresponding values at pH 14: it is given in Figure 8b. Notice that it includes manganese(II) and manganese(III) as the solid species, $Mn(OH)_2$ and $MnO(OH)$, respectively, introduced in Block 1. Like $Co(OH)_2$ and $Co(OH)_3$, for example, these are the forms more appropriate to alkaline conditions. Each of the couples in Figure 8 (a and b) is written in full in the S343 *Data Book*.

(a) Complete the diagram by calculating the value of $E^{\ominus}(MnO_4^{2-}(aq)\,|\,MnO_2(s))$ appropriate to a solution of pH 14.

(b) To what extent are the potential diagrams consistent with the following observations drawn from Block 1.

(i) Manganese(VI) disproportionates to manganese(IV) and manganese(VII) in acid solution, but is stable in alkali.

(ii) Acid solutions of $Mn^{2+}(aq)$ are *not* oxidised to $Mn^{3+}(aq)$ in air, whereas an alkaline suspension of $Mn(OH)_2$ is rapidly oxidised in air to a hydrated oxide of manganese(III), represented as MnO(OH).

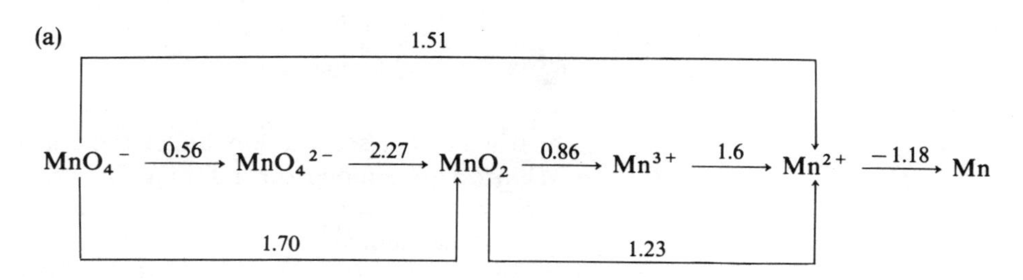

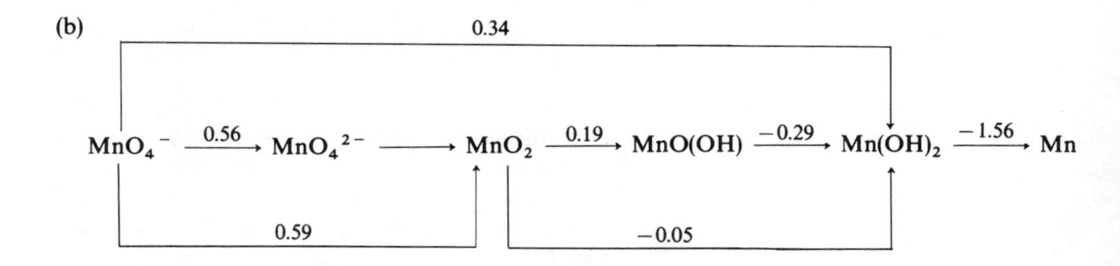

Figure 8 Potential diagrams for manganese in
(a) acid solution (pH 0), and
(b) alkaline solution (pH 14).

6 TRANSITION METALS AS PLANT MICRONUTRIENTS

The systems examined in Section 5 (and particularly in the final group of SAQs) are typical of the way in which the Nernst equation extends our grasp on the rich variety of transition-metal chemistry—at least in so far as redox processes in aqueous solution are concerned. Rather than work relentlessly through further examples taken from Block 1, we turn here to an important practical problem (familiar to gardeners!), which draws on one or two of the systems you have already analysed.

The chemical elements known to be essential for plant growth (see Figure 9, for example) include a group of transition metals, notably iron, manganese, zinc, copper and molybdenum. Only tiny concentrations of these elements are needed by the plant (usually less than 10^{-4} g per gram of dry matter), but they are essential (see Table 7, for example): hence they are called **micronutrients**, or **trace elements**. Deficiencies of the different micronutrients may arise for a variety of reasons—high rainfall leading to extensive leaching of light soils, for example, or overly intensive agriculture—but we shall concentrate here on the *effect of soil pH on the availability of iron.*

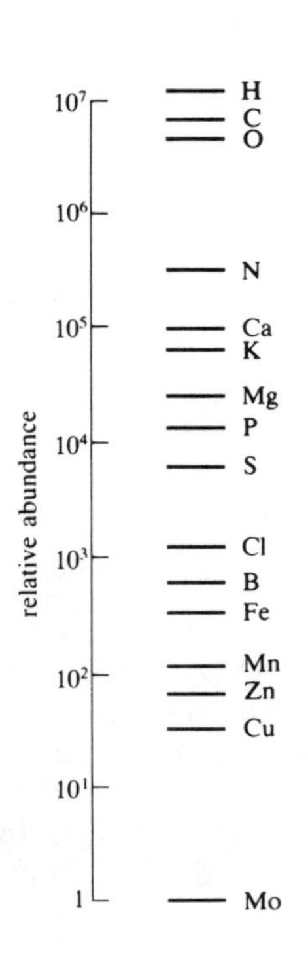

Figure 9 Relative abundance of the essential elements in alfalfa (at flowering stage). Cobalt is generally present in even smaller quantities than molybdenum in plants.

Table 7 Examples of the functions in higher plants of some micronutrients

Micronutrient	Typical functions
iron	redox catalysts (e.g. in respiration, photosynthesis, nitrogen-fixation, etc.); enzyme and protein constituent
manganese	nitrogen and inorganic acid metabolism; photosynthesis; enzyme activator
zinc	seed and grain maturation and production; formulation of growth hormone
copper	redox catalyst in respiration and photosynthesis; carbohydrate and protein metabolism
molybdenum	symbiotic nitrogen fixation and redox enzymes

For reasons that are not fully understood, iron deficiency impairs the production of chlorophyll, among other things. It is thus revealed by a yellowing of plant leaves called *chlorosis*, a condition that is particularly marked in sensitive species of the Rose family—certain fruit trees (see Plate 3.1 in *Colour Sheet 1*), for example—and in some ornamental plants, such as rhododendrons and other *Ericaceae*, when grown on soils of high pH. The underlying reason for this problem—and the most effective way of alleviating it—can be understood in terms of the ideas developed in Sections 4 and 5.

Thus, in working through SAQ 14 you established that aerial oxidation of iron(II) to iron(III) is thermodynamically favourable at all pH values, but it is *faster* under neutral or alkaline conditions than in acid. In other words, oxidation of iron(II) to iron(III) is favoured by a high pH, albeit because of kinetic rather than thermodynamic factors. Now, the pH of soils varies between about 4 and 10. At the lower end of this range, most of the iron will be present as iron(II)—whether generated by weathering from rocks, or through the bioreduction of iron(III) by microorganisms. The presence of iron(II) ensures a significant concentration of iron in the moisture associated with the soil, because even so-called 'insoluble' compounds of iron(II), such as $Fe(OH)_2$, have solubility products that are not too small (around 10^{-16}). But if a soil has a higher pH (in the range 7–10, say), conversion of iron(II) to iron(III) by aerial oxidation will be relatively fast—especially if the soil has an open, well-aerated texture. This ensures that the iron soon finishes up as a much more insoluble hydrated oxide of iron(III)—commonly known as rust! To simplify matters, we shall write this as $Fe(OH)_3$.

TLC 2

To get a feel for the problem this creates, suppose that a plant is nourished by a soil of pH 8, in which all iron is present as $Fe(OH)_3$—with a solubility product of about 10^{-39} (SAQ 14). Suppose further that the plant relies on the soluble iron present as $Fe^{3+}(aq)$ in the soil solution.

□ What is the maximum concentration of $Fe^{3+}(aq)$?

■ At pH 8, $a(H^+) = 10^{-8}$, so $a(OH^-) = 10^{-6}$. For $Fe(OH)_3$,

$$Fe(OH)_3(s) = Fe^{3+}(aq) + 3OH^-(aq) \qquad 62$$

$$K_{sp}^{\ominus} = a(Fe^{3+})\{a(OH^-)\}^3 = 10^{-39}$$

So $a(Fe^{3+}) = 10^{-39}/10^{-18} = 10^{-21}$.

No wonder high pH, combined with good aeration, can result in iron deficiency!

In principle, there are a number of ways of tackling this problem. The first, and simplest, is to add soluble iron compounds, such as $FeSO_4 . 7H_2O$. However, since such compounds are as subject to oxidation and precipitation as is naturally occurring iron(II), this approach is unlikely to be very efficient in practice. A second strategy is to lower the soil pH, thereby both increasing $a(Fe^{3+})$ (*cf.* the calculation above), and slowing down the oxidation of soluble iron(II). In practice, this is usually achieved by adding sulphur, which is oxidised to sulphate by the action of soil bacteria: the overall effect can be represented (in highly simplified form) as

$$S + \tfrac{3}{2}O_2 + H_2O \xrightarrow[\text{bacteria}]{\text{soil}} 2H^+ + SO_4^{2-} \qquad \mathbf{63}$$

Although effective (albeit on a fairly short-term basis), this is a rather drastic approach, since changing pH can (and does) affect both the supply of other nutrients, and the activity of soil organisms. Indeed, nutrients—especially micronutrients—can become 'too' available at low pH, reaching concentrations at which they become toxic to plants!

□ Can you suggest another way of increasing the concentration of soluble iron—without changing the pH?

■ Add to the soil a ligand that forms a stable, soluble complex of iron(III).

A reagent that forms an exceptionally stable complex with Fe^{3+}(aq) is the hexadentate ligand ethylenediaminetetraacetate ion, $edta^{4-}$. Thus, the stability constant is around 10^{25} for the reaction

$$Fe^{3+}(aq) + edta^{4-}(aq) = [Fe(edta)]^-(aq) \qquad \mathbf{64}$$

Recall now our soil solution at pH 8 in which equilibrium 62 maintains $a(Fe^{3+})$ at 10^{-21}. Suppose that $edta^{4-}$ is added such that $a(edta^{4-})$ is subsequently maintained at 10^{-5}.

□ What is $a\{[Fe(edta)]^-\}$ when both equilibria, 62 and 64, are satisfied?

■ Equilibrium 64 is satisfied by the condition:

$$\frac{a\{[Fe(edta)]^-\}}{a(Fe^{3+})a(edta^{4-})} = 10^{25}$$

With $a(Fe^{3+}) = 10^{-21}$ (which satisfies equilibrium 62 at pH 8) and $a(edta^{4-}) = 10^{-5}$, $a\{[Fe(edta)]^-\} = 10^{-1}$.

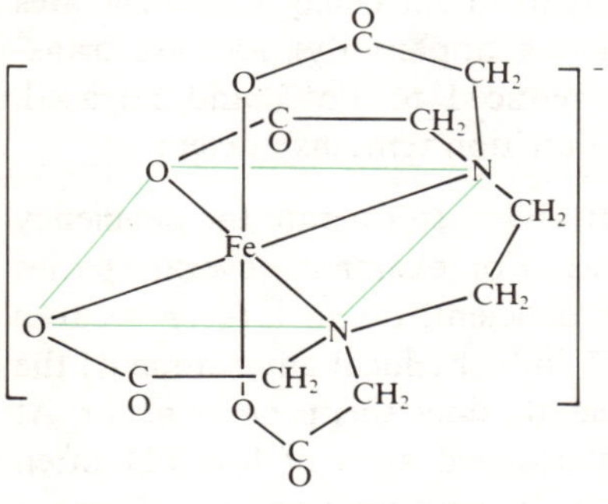

As suggested by this simple calculation, addition of the *chelating* ligand $edta^{4-}$ can drastically increase the maximum concentration of soluble iron in the soil solution. In a very real sense, the chelate structure, **1**, protects the iron(III) ion from the depredations of its watery environment—and hence effectively prevents the formation of rust.*

1

Although solutions containing free $edta^{4-}$(aq) (or other similar ligands) are sometimes used to rectify iron deficiency, it is more common to apply the Fe(III) complex—usually referred to as *'sequestered' iron*. Despite their expense, the use of such synthetic chelates is substantial—especially to meet micronutrient deficiencies of citrus and other fruit trees in many parts of the western United States, which have calcareous soils of high pH.

□ Look back at your answers to SAQs 15 and 16. Would you expect manganese deficiency to be a problem in well-aerated soils of relatively high pH?

■ Yes. Here, *thermodynamics* dictates that oxidation to manganese(III) (or to the even more insoluble +4 state) is favoured by high pH.

Manganese deficiency also causes chlorosis: as with iron, increasing soil acidity, adding $MnSO_4.4H_2O$ or manganese(II) chelates can all be effective. Indeed, chelates of other micronutrients, including zinc and copper, have also been used successfully in certain circumstances.

SLC 6

* It is interesting to note that nature has adopted much the same strategy in order to utilise iron at physiological pH in our own bodies. Thus, for example, the oxygen carriers *haemoglobin* and *myoglobin* both contain iron encircled by a tetradentate porphyrin ligand: the resulting haem is, in turn, encapsulated within a pocket in the globin protein. You will meet further examples later in the Course.

The mechanism by which micronutrients from chelates are absorbed by plants is still not fully understood, although it appears that the primary function of the chelate is indeed to keep the metal available in the soil solution, and hence able to diffuse to the root surface. Further, there is evidence from studies on certain types of plant that the iron is taken up as *uncomplexed* Fe^{2+}. Thus, the absorption process at the root surface must involve reduction and dissociation of the chelate—as suggested schematically in Figure 10. This releases the ligand for complexing with further Fe^{3+} in the soil particles—thereby enhancing the effectiveness of the initial application of chelate.

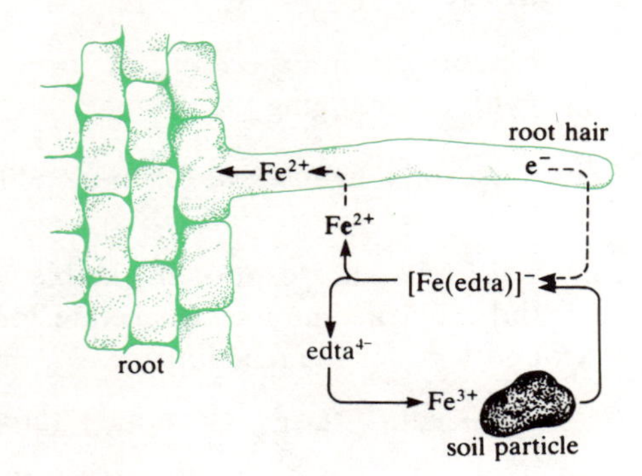

Figure 10 Schematic representation of the action of chelates (e.g. $edta^{4-}$) in solubilising and transporting iron from soil particle to root-hair surface.

Before leaving this Section, we should stress that the analysis outlined above presents a grossly oversimplified picture of the soil system. Thus, to suggest that soluble iron(III) in equilibrium with $Fe(OH)_3$ in the soil is in the form of $Fe^{3+}(aq)$ alone gives a misleading impression. Humus in the soil contains chelating ligands, which form complexes with $Fe^{3+}(aq)$*, thereby increasing the soluble iron. Addition of an artificial chelating agent only supplements the natural chelating powers of the soil—always providing, of course, that steps are taken to maintain the humus content. These natural chelating agents are not well characterised, but there is evidence that phenolic compounds, proteins, amino acids and organic acids all contribute. Further, there is a group of compounds—known collectively as *siderophores*—which are produced and released into the soil by many fungi and some bacteria. These compounds form extremely stable chelates with iron(III) (but not with iron(II) or other metals, apparently), and are transported into microbial cells where the iron is reduced to Fe^{2+} and released. However, their importance to overall plant iron nutrition remains unclear.

One final point: plants differ enormously in their ability to tolerate the deficiency of a given micronutrient—or, equally, an excess! For example, several species that absorb iron effectively even as it becomes deficient, do so largely because their roots are able to release reductants *and* H^+ into the local environment; the more deficient the conditions are, the more efficiently does this process occur. At the other end of the scale, poorly aerated waterlogged soils of low pH often contain concentrations of $Fe^{2+}(aq)$ that would be toxic to many plants. There is evidence that the roots of species tolerant of these conditions can *oxidise* the iron to the $+3$ state, and hence render it less available.

* ... and other metal cations. Indeed, the movement of iron, zinc, cobalt and other heavy metals in soils is thought to be associated with the formation of organic chelates.

7 COMPLEX FORMATION AND THE STABILITIES OF OXIDATION STATES—CONCLUDING REMARKS

As you have seen, the Nernst equation provides a quantitative framework within which to discuss the effects of complex formation on the relative stabilities of different oxidation states of a given metal. We have also touched on the chemical consequences of these effects, both in the laboratory, and in natural systems as diverse as the ocean bottom and the soil. But nowhere, so far, have we attempted to provide a theoretical explanation for these phenomena.

To return to the cobalt(III)/cobalt(II) system, for example, the values of $E^{\ominus}$ collected in Table 8 indicate that, like ammonia, the bidentate ligands ethylenediamine (en), 1,10-phenanthroline (phen) and 2,2′-bipyridyl (bipy) all stabilise cobalt(III) with respect to cobalt(II)—as does cyanide and the hexadentate $edta^{4-}$. Where such data exist for other metals of the first transition series, they suggest that this is a fairly common pattern; that is, complex formation usually (but not always, as we shall see) tends to stabilise the higher oxidation state. Again, according to the information in Table 8, the *degree* to which the ligand stabilises cobalt(III)—as measured by the move to an increasingly more negative value of $E^{\ominus}$—appears to follow the sequence: $H_2O < edta^{4-} < bipy \approx phen < NH_3 < en < CN^-$. Why should this be so? Is it, perhaps, possible that the theories expounded in Block 2 can help to rationalise data like these for a *given metal*—much as they did the trends in relative stability for a *given ligand* (H_2O, for example) *across* the first transition series (as outlined in Sections 2 and 3)?

Table 8 Standard electrode potentials at 298.15 K for selected Co^{III}/Co^{II} couples

Couple	$E^{\ominus}/V$
$Co^{3+}(aq) + e = Co^{2+}(aq)$	+1.94
$[Co(edta)]^{-}(aq) + e = [Co(edta)]^{2-}(aq)$	+0.38
$[Co(bipy)_3]^{3+}(aq) + e = [Co(bipy)_3]^{2+}(aq)$	+0.34
$[Co(phen)_3]^{3+}(aq) + e = [Co(phen)_3]^{2+}(aq)$	+0.33
$[Co(NH_3)_6]^{3+}(aq) + e = [Co(NH_3)_6]^{2+}(aq)$	+0.10
$[Co(en)_3]^{3+}(aq) + e = [Co(en)_3]^{2+}(aq)$	−0.18
$[Co(CN)_6]^{3-}(aq) + e = [Co(CN)_6]^{4-}(aq)$	−1.0

It turns out that questions like these are rather complicated and difficult to analyse, so much so that no complete, nor completely satisfactory, theoretical explanation exists as yet. Some understanding of the reasons for this can be gained by setting up a direct comparison with the cross-series problem, as tackled in Section 2. There, we began by drawing a thermodynamic cycle around the aqueous couple (Figure 2, Section 2.2), written as an oxidation:

$$M^{2+}(aq) + H^+(aq) = M^{3+}(aq) + \tfrac{1}{2}H_2(g) \qquad \mathbf{1}$$

whence we derived the following expression for $\Delta G_m^{\ominus}$ (*cf.* equation 7):

$$\Delta G_m^{\ominus}(1) = I_3(M) + \Delta H_h^{\ominus}(M^{3+}, g) - \Delta H_h^{\ominus}(M^{2+}, g) + \Delta H_H^{\ominus} - T\Delta S_m^{\ominus}(1) \qquad \mathbf{65}$$

If octahedral coordination is represented by attachment to 6L—irrespective of whether the ligand, L, is actually unidentate or bidentate (or whatever)—then an analogous cycle for the cobalt systems in Table 8 is the one given in Figure 11: for simplicity L is taken to be uncharged, but the arguments that follow clearly hold even when it is not.

$$[CoL_6]^{2+}(aq) + H^+(aq) \xrightarrow{\Delta H_m^{\ominus}(comp)} [CoL_6]^{3+}(aq) + \tfrac{1}{2}H_2(g) \qquad \mathbf{66}$$

$[CoL_6]^{2+}(aq) + H^+(aq)$ → (−6L(aq), $-\Delta H_L^{\ominus}(Co^{2+}, g)$) → $Co^{2+}(g) + H^+(aq)$ → ($I_3 + \Delta H_H^{\ominus}$) → $Co^{3+}(g) + \tfrac{1}{2}H_2(g)$ → (+6L(aq), $\Delta H_L^{\ominus}(Co^{3+}, g)$) → $[CoL_6]^{3+}(aq) + \tfrac{1}{2}H_2(g)$

Figure 11 Thermodynamic cycle around equation 66.

Here, $\Delta H_L^{\ominus}(Co^{3+}, g)$, for example, is the standard enthalpy of immersion of the gaseous $Co^{3+}(g)$ ion in an aqueous solution of the ligand, with formation of the hydrated complex, $[CoL_6]^{3+}(aq)$—the process represented by the right-hand vertical arrow:

$$Co^{3+}(g) + 6L(aq) = [CoL_6]^{3+}(aq); \quad \Delta H_m^{\ominus} = \Delta H_L^{\ominus}(Co^{3+}, g) \qquad \textbf{67}$$

This, and the corresponding term $\Delta H_L^{\ominus}(Co^{2+}, g)$ are thus analogous to the more familiar enthalpies of hydration, $\Delta H_h^{\ominus}(M^{3+}, g)$ and $\Delta H_h^{\ominus}(M^{2+}, g)$, respectively, in equation 65.

As you may wish to check for yourself, this cycle leads to the following expression for $\Delta G_m^{\ominus}(comp)$, the value of $\Delta G_m^{\ominus}$ for the cobalt couple as written in equation 66:

$$\begin{aligned}\Delta G_m^{\ominus}(comp) &= \Delta H_m^{\ominus}(comp) - T\Delta S_m^{\ominus}(comp)\\ &= I_3(Co) + \Delta H_L^{\ominus}(Co^{3+}, g) - \Delta H_L^{\ominus}(Co^{2+}, g)\\ &\quad + \Delta H_H^{\ominus} - T\Delta S_m^{\ominus}(comp)\end{aligned} \qquad \textbf{68}$$

□ How is $\Delta G_m^{\ominus}(comp)$ related to $E_{comp}^{\ominus}$ as defined in Section 5.5?

■ $\Delta G_m^{\ominus}(comp) = +FE_{comp}^{\ominus}$; $E_{comp}^{\ominus}$ is the standard potential of the complex-ion couple written as a *reduction*: for cobalt,

$$[CoL_6]^{3+}(aq) + e = [CoL_6]^{2+}(aq) \qquad \textbf{69}$$

Recalling that e represents $\{\frac{1}{2}H_2(g) - H^+(aq)\}$, equation 69 is the *reverse* of equation 66, so $\Delta G_m^{\ominus}(comp) = -\Delta G_m^{\ominus}(69) = -(-nFE_{comp}^{\ominus}) = +FE_{comp}^{\ominus}$ (since $n = 1$ here).

The formal similarity between equations 65 and 68 belies important differences between them once attention shifts to our present problem. To begin with, fixing the metal fixes the value of I_3. Since $\Delta H_H^{\ominus}$ is also constant (Section 2.2), this means that *changes* in $\Delta G_m^{\ominus}(comp)$ with the ligand L can be discussed in terms of a simplified version of equation 68, as:

$$\Delta G_m^{\ominus}(comp) = \Delta H_L^{\ominus}(Co^{3+}, g) - \Delta H_L^{\ominus}(Co^{2+}, g) - T\Delta S_m^{\ominus}(comp) + \text{constant} \qquad \textbf{70}$$

However, this apparent simplification is more than offset by the problem of discerning a clear-cut pattern in the contributions of the remaining quantities. There are two major difficulties.

Firstly, we cannot, in general, ignore the entropy term in equation 70. Recall that earlier we were concerned with variations across a series of closely similar—or 'analogous'—reactions. Thus, we had grounds (backed by experimental evidence in the case of reaction 1) for assuming that the entropy term had a roughly constant effect across the series. The present problem is very different. Reliable thermodynamic data for 'families' of systems like that in Table 8 are very scanty, but there is no *a priori* justification for assuming that changes in $\Delta S_m^{\ominus}(comp)$ from ligand to ligand will be relatively small. Indeed, such data as do exist suggest that this assumption may sometimes rest on rather shaky ground—especially for systems where 'solvation effects' (linked to the hydration of ionic species) are likely to differ markedly when the ligand is changed. The entropy is known to be particularly sensitive to such changes (witness the example in SAQ 17 below).

Secondly, it is usually much more difficult to explain variations in the enthalpy terms—$\Delta H_L^{\ominus}(Co^{2+}, g)$ and $\Delta H_L^{\ominus}(Co^{3+}, g)$ in our cobalt example—with the ligand L, than it is to explain variations in the corresponding terms—$\Delta H_h^{\ominus}(M^{2+}, g)$ and $\Delta H_h^{\ominus}(M^{3+}, g)$, for example—with the metal M. This is essentially because the metal ions (and hence their interactions with H_2O ligands) change in a regular and intelligible way across the transition series: in particular, in crossing the series, successive protons are added to the nucleus, and successive electrons to the 3d shell. By contrast, arbitrary changes in the ligand combination L_6 from, say, $(H_2O)_6$ to $(edta^{4-})$ to $(phen)_3$ show no such regularities. (Refer again to Exercise 1 and its answer if need be.)

Despite the difficulties outlined above, it is worth pursuing the analysis a little, in order to highlight one or two of the more important factors 'at work', as it were, in the Co^{III}/Co^{II} systems. To take a cue from our analysis of the cross-series problem: recall that the variation in $\Delta G_m^\ominus$ for reactions like that in equation 1 was often dominated by a single term, namely I_3. Moreover, deviations from this pattern could then be attributed (in a qualitative way, at least) to enhanced contributions from one or more of the other terms involved (Section 3.1). Perhaps there is an equivalent term this time?

Suppose we start by neglecting changes in the entropy term. Then we are left with variations in $\Delta G_m^\ominus$(comp) being attributed to a *difference* between two enthalpy terms, each of which effectively represents the energy of an interaction in which a gaseous metal ion (either Co^{3+} or Co^{2+}) surrounds itself by six ligands.

□ In this simplified view, when will complexation with a given ligand lead to stabilisation of the +3 state as compared with the aquo couple, M^{3+}(aq)/ M^{2+}(aq)?

■ As summarised in Section 5.5, if $E_{comp}^\ominus$ is more negative than $E^\ominus(Co^{3+} | Co^{2+})$, then the ligand stabilises the higher oxidation state. Since $\Delta G_m^\ominus(\text{comp}) = +FE_{comp}^\ominus$, this translates into a more negative value of $\Delta G_m^\ominus$(comp). In other words, we require the difference $\{\Delta H_L^\ominus(Co^{3+}, g) - \Delta H_L^\ominus(Co^{2+}, g)\}$ to be *more negative* when the ligand is L than when it is H_2O.

Now, according to ligand-field theory, terms like $\Delta H_L^\ominus(Co^{3+}, g)$ and $\Delta H_L^\ominus(Co^{2+}, g)$ have two parts—one due to simple electrostatic interaction between the metal ion and the ligands surrounding it, and the second (the ligand-field stabilisation energy, LFSE) due to the ligand-field splitting of the 3d orbitals. For a *given* ligand, the electrostatic part will always favour the higher oxidation state, by virtue of the higher charge on the Co^{3+} ion. However, the way this interaction—and more importantly, its contribution to the difference $\{\Delta H_L^\ominus(Co^{3+}, g) - \Delta H_L^\ominus(Co^{2+}, g)\}$—varies from ligand to ligand is more difficult to predict.

On the other hand, the LFSE contribution *does* depend explicitly on the nature of the ligand, which, together with the oxidation state of the metal, determines the size of the splitting, Δ_o. Suppose we concentrate on this contribution.

□ With the exception of $[Co(CN)_6]^{4-}$(aq), all of the cobalt(II) complexes in Table 8 are high spin. What is the LFSE for such a complex? (Refer to Block 2 if necessary.)

■ Cobalt(II) has the configuration d^7, so the LFSE in a high-spin octahedral complex is $\frac{4}{5}\Delta_o$.

By contrast, *all* the corresponding cobalt(III) complexes are low-spin d^6, so that here the LFSE is $(\frac{12}{5}\Delta_o - 2P)$. Assuming that the pairing energy P does not change from ligand to ligand, then the variation in the LFSE contribution is determined by $\frac{12}{5}\Delta_o$—the *maximum conceivable amount* for any set of complexes. Given that the value of Δ_o for any particular ligand will also be greater in the +3 oxidation state, it seems that we may well have lighted upon the 'dominant' term we were after.

□ If this LFSE part of $\Delta H_L^\ominus(Co^{3+}, g)$ does dominate in the way $\Delta G_m^\ominus$(comp) varies with L, what should determine the degree to which a particular ligand stabilises the +3 state?

■ If this approximation holds, then the variation in $\Delta G_m^\ominus$(comp) should reflect the relative size of Δ_o for the ligand in question. A measure of the latter is, in turn, provided by the *position* of the ligand in the spectrochemical series:

$$I^- < Br^- < Cl^- < H_2O < edta^{4-} < NH_3 < en < bipy < phen < CN^-$$

We would then expect $E^{\ominus}_{\text{comp}}$ for the cobalt systems in Table 8 to vary as: $H_2O > edta^{4-} > NH_3 >$ en > bipy > phen > CN^-, compared with the observed sequence: $H_2O > edta^{4-} >$ bipy ≈ phen > NH_3 > en > CN^-. This is a remarkable level of agreement, given the many assumptions and approximations inherent in our analysis.

Now try the following SAQ.

SAQ 17 Table 9 contains $E^{\ominus}$ values for a number of iron(III)/iron(II) couples, together with corresponding values of $\Delta H^{\ominus}_m$(comp) and $\Delta S^{\ominus}_m$(comp) for the couples written as oxidations, that is (*cf.* equation 66):

$$[FeL_6]^{2+}(aq) + H^+(aq) = [FeL_6]^{3+}(aq) + \tfrac{1}{2}H_2(g) \quad \textbf{71}$$

With the exception of the aquo couple, all of the other complexes are low spin. Discuss briefly the extent to which the observed effects of complex formation tie in with the simplified analysis outlined in this Section.

Table 9 Standard electrode potentials at 298.15 K for selected Fe^{III}/Fe^{II} couples, together with corresponding values of $\Delta H^{\ominus}_m$(comp) and $\Delta S^{\ominus}_m$(comp)

Couple	$E^{\ominus}$/V	$\Delta H^{\ominus}_m$/kJ mol^{-1}	$\Delta S^{\ominus}_m$/J K^{-1} mol^{-1}
$Fe^{3+}(aq) + e = Fe^{2+}(aq)$	+0.77	40.6	−113.4
$[Fe(bipy)_3]^{3+}(aq) + e = [Fe(bipy)_3]^{2+}(aq)$	+1.11	136.8	99.6
$[Fe(phen)_3]^{3+}(aq) + e = [Fe(phen)_3]^{2+}(aq)$	+1.13	137.8	87.0
$[Fe(CN)_6]^{3-}(aq) + e = [Fe(CN)_6]^{4-}(aq)$	+0.36	106.3	240.6

The example in SAQ 17 provides a salutary reminder of the limitations of the theoretical approach presented above. It should now be apparent that we chose our first example with care—to provide a sort of 'best case'. Thus, in the cobalt systems, all of the 'energy factors' considered appear to act together—in the direction of increasing stability of the +3 oxidation state with increasing Δ_o of the ligand: only phen and bipy seem to be out of sequence. At the same time, any adverse variations in the entropy term are apparently not dramatic enough to make themselves felt.

By contrast, once the factors identified here begin to 'fight against' one another—as in the iron systems—it becomes much more difficult to predict the outcome with any confidence. Indeed, it must be admitted that a satisfactory theoretical explanation of such cases does not exist. Nevertheless, you will see later in the Course that the somewhat fuller picture provided by molecular-orbital theory (as introduced in Block 2) can provide a qualitative explanation for some of the additional energy factors involved. A notable example is the way π-bonding with certain strong-field ligands—like phen and bipy, for instance—can preferentially stabilise *low* oxidation states. It is tempting to speculate that this effect may be responsible for displacing complexes with these two ligands from their 'expected' position in the family of cobalt systems.

Given difficulties like the ones rehearsed above, it is not surprising that workers in the field have, in fact, tended to adopt a different, more empirically based approach to explaining the stability of metal complexes—one that is not restricted to transition metals. This subject is taken up later in the Course. Suffice it to say here that it rests on a number of quite powerful generalisations drawn from the vast literature on metal complexes and their stability constants. As you will see, this approach has proved to be remarkably successful in rationalising the selection of ligands to stabilise different oxidation states of a given metal.

OBJECTIVES FOR BLOCK 3

Now that you have completed Block 3, you should be able to do the following things:

1 Recognise valid definitions of, and use in a correct context, the terms, concepts and principles in Table A.

Table A List of scientific terms, concepts and principles used in Block 3

Term	Page No.
activity, a	18
enthalpy of hydration, $\Delta H_h^{\ominus}$	5
entropy change for analogous reactions	6
exchange energy	*AV Booklet*
'hydrogen electrode'	24
micronutrient (trace element)	27
Nernst equation	18
'oxygen electrode'	23
pH	23
solubility product, K_{sp}	16
stability constant	15
standard (thermodynamic) equilibrium constant $K^{\ominus}$	20

2 Show how the interplay of the quantities in an appropriate thermodynamic cycle leads to a correlation between the variations in the relative stabilities of the M^{2+}(aq) and M^{3+}(aq) ions, the variation in the relative stabilities of the dihalides and trihalides, and the variation in the third ionisation energies of the first-row transition elements. (Exercise 2)

3 Recognise imperfections in the correlation described in Objective 2, and account for them in terms of ligand-field stabilisation energies. (SAQs 1 and 6)

4 Give examples which show that the correlations described in Objective 2 are apparent in the chemical behaviour of the compounds concerned. (SAQs 4 and 5)

5 Relate the variation in the third ionisation energies of the first-row transition metals to nuclear charge, classical electron repulsion and exchange energy. (SAQ 2)

6 Define the activity of: (a) a solid or liquid; (b) an ideal gas; and (c) an ideal solute in solution. (SAQs 10–12, 14 and 15)

7 Write down the standard equilibrium constant $K^{\ominus}$ of a given reaction, and use the definitions in Objective 6 to relate this to the corresponding 'experimental' equilibrium constant. (SAQs 7, 8 and 11)

8 Use the Nernst equation to determine the 'concentration' dependence and/or pH dependence of individual redox potentials. (SAQs 10, 12–16).

9 Relate changes in the stabilities of oxidation states brought about by complexing or precipitation to the relative sizes of stability constants or solubility products:

either (a) directly; (SAQ 9)

or (b) by using the procedure described in Objective 8 to calculate and compare appropriate redox potentials. (SAQs 12, 13, 14, 16 and 17)

10 Use appropriate redox potentials to determine whether an aqueous ion in an aerated solution of given pH is:

(a) thermodynamically unstable with respect to reduction by the solvent system;

(b) thermodynamically unstable with respect to oxidation by the solvent system;

(c) thermodynamically unstable with respect to disproportionation;

(d) thermodynamically unstable with respect to oxidation/reduction by any other reagent. (SAQs 13, 14, 15 and 16)

11 State the limitations of the predictions outlined in Objectives 9 and 10. (SAQs 13, 14 and 16)

12 Given appropriate data, comment on the effects of complex formation on the relative stability of oxidation states of a given metal in terms of ligand-field theory. (SAQ 17)

SAQ ANSWERS AND COMMENTS

SAQ 1
(Objective 3)

Figure 3 suggests that the departure from parallelism is due to a comparatively large contribution from the term $\{\Delta H_h^{\ominus}(M^{3+}, g) - \Delta H_h^{\ominus}(M^{2+}, g)\}$. Thus, from vanadium to chromium there is a large decrease in $\Delta H_h^{\ominus}(M^{3+}, g)$, whereas $-\Delta H_h^{\ominus}(M^{2+}, g)$ is virtually unchanged.

SAQ 2
(Objective 5)

From boron to neon, the electronic configuration changes from $1s^22s^22p^1$ to $1s^22s^22p^6$, so we may consider the outer 2p electrons outside a core composed of the nucleus and the filled 1s and 2s shells—a core with the configuration of beryllium, in other words. The calculation of the ionisation energy follows the analysis in the AV sequence, leading to an equation analogous to equation 11:

$$E(p^{n-1}) - E(p^n) = U - (n-1)J + K\,\delta m$$

Values of δm for this system are collected in Table 10: from p^1 to p^3, δm increases from 0 to 2; it drops to zero again at p^4, before repeating the increase up to 2 at p^6. If this variation is imposed on a smoothly increasing function $\{U - (n-1)J\}$, the plot in Figure 5 results.

Table 10 The decrease in the number of pairs of parallel spins on ionisation for electronic configurations p^1 to p^6

Ionisation process	δm
$p^1 \rightarrow p^0$	0
$p^2 \rightarrow p^1$	1
$p^3 \rightarrow p^2$	2
$p^4 \rightarrow p^3$	0
$p^5 \rightarrow p^4$	1
$p^6 \rightarrow p^5$	2

SAQ 3
(revision)

The simplest way of establishing a connection is to write the disproportionation reaction, equation 14, as the sum of the following reactions:

$$2ScCl_2(s) + Cl_2(g) = 2ScCl_3(s) \qquad \mathbf{72}$$

$$ScCl_2(s) = Sc(s) + Cl_2(g) \qquad \mathbf{73}$$

$$3ScCl_2(s) = Sc(s) + 2ScCl_3(s) \qquad \mathbf{14}$$

Then $\Delta G_m^{\ominus}(14) = \Delta G_m^{\ominus}(72) + \Delta G_m^{\ominus}(73)$. Thus, the more negative the value of $\Delta G_m^{\ominus}(72)$ $(= 2 \times \Delta G_m^{\ominus}(13))$, the more negative is $\Delta G_m^{\ominus}(14)$, and so the more unstable to disproportionation is $ScCl_2(s)$.

SAQ 4
(Objective 4)

At the beginning of the series, I_3 is low and MF_2 is unstable with respect to MF_3 (and M). Thus, ScF_2 and TiF_2 are unknown. At the end of the series, I_3 is high and MF_3 is unstable with respect to MF_2 (and $\frac{1}{2}F_2$). Thus, NiF_3, CuF_3 and ZnF_3 are unknown. Notice that because MnF_3 and FeF_3 are both known, the drop in I_3 from manganese to iron is not evident in the data on fluorides in Tables 2 and 3. Nevertheless, the following known values of $\Delta G_m^{\ominus}$ confirm the expected trend:

$$MnF_2(s) + \tfrac{1}{2}F_2(g) = MnF_3(s); \quad \Delta G_m^{\ominus} = -184\,\text{kJ mol}^{-1}$$

$$FeF_2(s) + \tfrac{1}{2}F_2(g) = FeF_3(s); \quad \Delta G_m^{\ominus} = -303\,\text{kJ mol}^{-1}$$

SAQ 5
(Objective 4)

(a) Yes; (b) no. $MnBr_3$ is unknown, but $FeBr_3$ is stable with respect to $FeBr_2$ and $\frac{1}{2}Br_2$ at normal temperatures. In other words, $\Delta G_m^{\ominus}$ for the reaction

$$MBr_2 + \tfrac{1}{2}Br_2 = MBr_3$$

correlates with I_3 and falls from manganese to iron: the value for Mn seems to be positive, but that for iron is negative, so $MnBr_3$ is unstable to dissociation, whereas $FeBr_3$ is stable. For the iodides, both $\Delta G_m^{\ominus}$ values must be positive since both tri-iodides are unstable. Nevertheless, our correlation suggests that the value for iron must be less positive than that for manganese.

SAQ 6
(Objective 3)

The more negative the redox potential quoted in Table 4, the more negative is the value of $\Delta G_m^{\ominus}$ for reaction 15. Thus, from chromium to cobalt, $\Delta G_m^{\ominus}$ for equation 15 varies in the sequence: $Cr < Mn < Fe > Co$. This sequence departs noticeably from that in I_3 $(Cr < Mn > Fe < Co)$: in particular, the downward break in $\Delta G_m^{\ominus}$ occurs from iron to cobalt, rather than from manganese to iron.

SAQ 7
(revision and Objective 7)

The form of the equilibrium constant expression follows directly from the stoichiometry of the balanced reaction equation.

(a) The required expression is straightforward:

$$K = \frac{c\{[Co(NH_3)_6]^{3+}\}}{c(Co^{3+})\{c(NH_3)\}^6}$$

(b) and (c) Here, the questions arise as to what is meant by the 'concentration' of a gas (equation 27), or a solid (equation 28), respectively. As you may know from other courses, in this context the concentration of a gas is measured by its pressure, which leads to the following expression for the equilibrium constant of reaction 27:

$$K = \frac{\{c(Co^{2+})\}^2\{c(H^+)\}^2}{\{c(Co^{3+})\}^2 p(H_2)}$$

where $p(H_2)$ represents the pressure of hydrogen.

On the other hand, a solid is taken to have a 'constant' effect, which can be 'incorporated' into the equilibrium constant. For reaction 28, this suggests the following expression:

$$K = c(Co^{2+})\{c(OH^-)\}^2$$

A deeper explanation for these apparently quite arbitrary conventions is taken up in Section 5.3.

SAQ 8
(revision and Objective 7)

For reaction 28,

$$\begin{aligned} \Delta G_m^\ominus &= \Delta G_m^\ominus(Co^{2+}, aq) + 2\Delta G_m^\ominus(OH^-, aq) - \Delta G_m^\ominus(Co(OH)_2, s) \\ &= \{-54.5 + 2(-157.2) - (-454.3)\}\,kJ\,mol^{-1} \\ &= 85.4\,kJ\,mol^{-1} \end{aligned}$$

According to equation 22,

$$\log K = \frac{-85.4 \times 10^3\,J\,mol^{-1}}{2.303 \times (8.314\,J\,K^{-1}\,mol^{-1}) \times (298.15\,K)} = -14.959\,6^*$$

Taking the 'inverse' logarithm to the base ten (which may be marked 10^x on your calculator), $K = 1.098 \times 10^{-15}$.

From the answer to part (c) of SAQ 7, the equilibrium constant for reaction 28 has exactly the same form as the solubility product, K_{sp}, of $Co(OH)_2$. The difficulty is that K as calculated from thermodynamic data is dimensionless, whereas K_{sp} (as defined in equation 29) has the dimension (concentration)^3, and the SI unit $mol^3\,l^{-3}$. This problem is also taken up in Section 5.3.

SAQ 9
(Objective 9)

The solubility product of $Co(OH)_2$ ($10^{-15}\,mol^3\,l^{-3}$) indicates that this is, indeed, a sparingly soluble compound. Not surprisingly, therefore, adding hydroxide to the system in equation 18, precipitates Co^{2+}(aq) as $Co(OH)_2$: it also lowers H^+(aq). Both of these effects tend to shift the equilibrium to the right-hand side, thus stabilising cobalt(II).

However, $Co(OH)_3$ is also sparingly soluble—and very much *less* soluble than $Co(OH)_2$. Precipitation of Co^{3+}(aq) as $Co(OH)_3$ evidently outweighs the effect of both precipitation of Co^{2+}(aq) as $Co(OH)_2$ and the lowering of H^+(aq); the net effect is to shift the equilibrium in equation 18 to the left, in accord with the observed behaviour of the system.

SAQ 10
(Objectives 6 and 8)

(a) Applying the Nernst equation to the Co^{3+}/Co^{2+} couple gives (since $n = 1$):

$$E = E^\ominus - \left(\frac{2.303RT}{F}\right)\log\left\{\frac{a(Co^{2+})}{a(Co^{3+})}\right\}$$

For aqueous solutes, Case 3 applies, so $a(Co^{2+}) = c(Co^{2+})/c^\ominus$ and $a(Co^{3+}) = c(Co^{3+})/c^\ominus$, and the equation above becomes:

$$E = E^\ominus - \left(\frac{2.303RT}{F}\right)\log\left\{\frac{c(Co^{2+})}{c(Co^{3+})}\right\}$$

* To help you check your calculations, we shall sometimes (as here) quote the result of an intermediate step. Any such result will be quoted to sufficient significant figures to ensure that the final answer is correct (within the accuracy of the original data). Make sure you do the same! In other words, never 'round off' an intermediate result—only the final answer.

(b) Here, $n = 2$, so the Nernst equation gives:

$$E = E^{\ominus} - \left(\frac{2.303RT}{2F}\right) \log \left\{\frac{a(H_2O)}{\{a(O_2)\}^{1/2}\{a(H^+)\}^2}\right\}$$

According to the definitions in Cases 1, 2 and 3, respectively, $a(H_2O) = 1$; $a(O_2) = p(O_2)/p^{\ominus}$; $a(H^+) = c(H^+)/c^{\ominus}$, and the equation becomes

$$E = E^{\ominus} - \left(\frac{2.303RT}{2F}\right) \log \left\{\frac{1}{\{p(O_2)/p^{\ominus}\}^{1/2}\{c(H^+)/c^{\ominus}\}^2}\right\}$$

At 298.15 K, the constant term in the Nernst equation is:

$$\frac{2.303RT}{nF} = \frac{2.303 \times (8.314\,\mathrm{J\,K^{-1}\,mol^{-1}}) \times (298.15\,\mathrm{K})}{n(96\,485\,\mathrm{C\,mol^{-1}})}$$

$$= (0.059\,2/n)\,\mathrm{J\,C^{-1}}$$

$$= (0.059\,2/n)\,\mathrm{V}$$

SAQ 11
(Objectives 6 and 7)

(a) Here, $K_{sp} = \{c(Co^{3+})\}\{c(OH^-)\}^3$

whereas, $K^{\ominus} = \dfrac{a(Co^{3+})\{a(OH^-)\}^3}{a(Co(OH)_3)}$

$$= \{c(Co^{3+})/c^{\ominus}\}\{c(OH^-)/c^{\ominus}\}^3$$

Comparing the two expressions reveals that

$$K^{\ominus} = K_{sp}/(c^{\ominus})^4 = 10^{-45} \text{ (provided } c^{\ominus} = 1\,\mathrm{mol\,l^{-1}}).$$

As for the examples in the text, $K^{\ominus}$ is obtained by simply striking out the units.

(b) Here, $K = \{p(NO_2)\}^2/p(N_2O_4)$

whereas, $K^{\ominus} = \{a(NO_2)\}^2/a(N_2O_4)$

$$= \{p(NO_2)/p^{\ominus}\}^2/\{p(N_2O_4)/p^{\ominus}\}$$

$$= (\{p(NO_2)\}^2/p(N_2O_4)) \times (1/p^{\ominus})$$

Comparing the two expressions reveals that $K^{\ominus} = K/p^{\ominus}$. Here, $K = 1.5 \times 10^4\,\mathrm{Pa} = 0.15\,\mathrm{bar}$, so $K^{\ominus} = 0.15$ (since $p^{\ominus} = 1\,\mathrm{bar}$).

Provided pressures are first converted to bars, $K^{\ominus}$ is again obtained by simply striking out the units.

SAQ 12
(Objectives 6, 8 and 9)

For $Co(OH)_2$ (see Section 5.3):

$$K^{\ominus} = \{c(Co^{2+})/c^{\ominus}\}\{c(OH^-)/c^{\ominus}\}^2 = 10^{-15}$$

At unit activity of OH^-, $\{c(Co^{2+})/c^{\ominus}\} = K^{\ominus} = 10^{-15}$. Likewise, for $Co(OH)_3$ under these conditions, $\{c(Co^{3+})/c^{\ominus}\} = K^{\ominus} = 10^{-45}$. Thus, the concentrations of Co^{2+}(aq) and Co^{3+}(aq) are $10^{-15}\,\mathrm{mol\,l^{-1}}$ and $10^{-45}\,\mathrm{mol\,l^{-1}}$, respectively.

According to the analysis in Section 5.4.1, $E^{\ominus}$ for the couple in equation 47:

$$Co(OH)_3(s) + e = Co(OH)_2(s) + OH^-(aq) \qquad \textbf{47}$$

is the value of E for the $Co^{3+}(aq)/Co^{2+}(aq)$ couple in a solution containing OH^-(aq) at unit activity, such that the concentrations of Co^{3+}(aq) and Co^{2+}(aq) are determined by the solubility products of $Co(OH)_3$ and $Co(OH)_2$, respectively, as calculated above. Thus, the Nernst equation for the aquo couple is (equation 42):

$$E = 1.94\,\mathrm{V} - (0.059\,2\,\mathrm{V}) \log \{c(Co^{2+})/c(Co^{3+})\}$$

$$= 1.94\,\mathrm{V} - (0.059\,2\,\mathrm{V}) \log \{10^{-15}/10^{-45}\}$$

$$= 0.16\,\mathrm{V}$$

which is $E^{\ominus}$ for the hydroxide couple in equation 47.

SAQ 13
(Objectives 8, 9, 10 and 11)

(a) For the oxygen electrode at pH 10, $E = (1.23 - 0.0592 \times 10)\,\text{V} = 0.64\,\text{V}$.

(b) Applying the Nernst equation to the hydroxide couple gives:

$$E = E^{\ominus} - (0.0592\,\text{V})\log\left\{\frac{a(\text{Co(OH)}_2)a(\text{OH}^-)}{a(\text{Co(OH)}_3)}\right\}$$

Since the activities of the two solids are both unity and $E^{\ominus} = 0.16\,\text{V}$, this reduces to:

$$E = 0.16\,\text{V} - (0.0592\,\text{V})\log a(\text{OH}^-)$$

at pH 10, $a(\text{H}^+) = 10^{-10}$, so $a(\text{OH}^-) = 10^{-14}/10^{-10} = 10^{-4}$, and

$$E = 0.16\,\text{V} - (0.0592\,\text{V})(-4) = 0.40\,\text{V}$$

Since the potential of the oxygen electrode is larger (more positive) than the corresponding value for the cobalt couple, aerial oxidation of $Co(OH)_2$ to $Co(OH)_3$ is thermodynamically favourable under these conditions. Since the reaction actually happens in practice, it must also be kinetically favourable.

SAQ 14
(Objectives 6, 8, 9, 10 and 11)

(a) The ratio of solubility products is 1.57×10^{-23}, giving $K^{\ominus}_{\text{sp}}\{\text{Fe(OH)}_3\}$ of about 10^{-39}. The Fe^{III}/Fe^{II} system is directly analogous to the cobalt system analysed in SAQ 12. Thus, $E^{\ominus}$ for the hydroxide couple (equation 58) is again the value of E for the $Fe^{3+}(aq)/Fe^{2+}(aq)$ couple in a solution containing $OH^-(aq)$ at unit activity.

Applying the Nernst equation to the aquo couple (equation 57) gives:

$$E = E^{\ominus} - (0.0592\,\text{V})\log\{a(\text{Fe}^{2+})/a(\text{Fe}^{3+})\}$$

$$\text{So}\log\{a(\text{Fe}^{2+})/a(\text{Fe}^{3+})\} = \left\{\frac{(-0.58\,\text{V}) - (0.77\,\text{V})}{-0.0592\,\text{V}}\right\} = +22.8041$$

whence $a(\text{Fe}^{2+})/a(\text{Fe}^{3+}) = 6.369 \times 10^{22}$.

This figure represents the ratio of activities of the aqueous ions in equilibrium with the solid hydroxides. Again, drawing on the answer to SAQ 12, for $Fe(OH)_2$:

$$K^{\ominus}_{\text{sp}}(\text{Fe(OH)}_2) = a(\text{Fe}^{2+})\{a(\text{OH}^-)\}^2$$
$$= a(\text{Fe}^{2+}) \text{ if } a(\text{OH}^-) = 1$$

Likewise, $a(\text{Fe}^{3+}) = K^{\ominus}_{\text{sp}}(\text{Fe(OH)}_3)$ under these conditions.

Thus $K^{\ominus}_{\text{sp}}(\text{Fe(OH)}_2)/K^{\ominus}_{\text{sp}}(\text{Fe(OH)}_3) = a(\text{Fe}^{2+})/a(\text{Fe}^{3+})$
$= 6.369 \times 10^{22}$

Hence the required ratio $= 1/(6.369 \times 10^{22}) = 1.57 \times 10^{-23}$.

Then $K^{\ominus}_{\text{sp}}\{\text{Fe(OH)}_3\} = 10^{-16} \times 1.57 \times 10^{-23} \approx 10^{-39}$.

(b) Making the solution alkaline leads initially to a precipitate of $Fe(OH)_2$ ($K^{\ominus}_{\text{sp}} = 10^{-16}$). Assuming pH = 14 in the resulting solution (for want of more specific information), the potential of the oxygen electrode is $E = (1.23 - 0.0592 \times 14) = 0.40\,\text{V}$. This is more positive than the corresponding value for the iron couple (equation 58), so aerial oxidation of $Fe(OH)_2$ to $Fe(OH)_3$ is thermodynamically favourable in alkaline solution as, indeed, is the corresponding oxidation of $Fe^{2+}(aq)$ to $Fe^{3+}(aq)$ in acid. In other words, oxygen is thermodynamically capable of oxidising iron(II) to iron(III) at all values of pH. The crucial point is that the process is very much *faster* in alkali (for test-tube quantities it takes about 5 minutes) than in acid, so here the observed 'stabilisation' of iron(III) with increasing pH rests on *kinetic* rather than thermodynamic factors.

By contrast, in the cobalt case, *thermodynamic* effects are crucial, because cobalt(II) is thermodynamically favoured in acid and cobalt(III) in alkali.

SAQ 15
(Objectives 6, 8 and 10)

Equation 60 is a combination of the following couples:

$$\text{MnO}_2(\text{s}) + 4\text{H}^+(\text{aq}) + 2e = \text{Mn}^{2+}(\text{aq}) + 2\text{H}_2\text{O(l)};\quad E^{\ominus} = 1.23\,\text{V}$$

$$\tfrac{1}{2}\text{O}_2(\text{g}) + 2\text{H}^+(\text{aq}) + 2e = \text{H}_2\text{O(l)};\quad E^{\ominus} = 1.23\,\text{V}$$

Applying the Nernst equation to the manganese couple gives:

$$E = E^{\ominus} - \left(\frac{0.0592\,\text{V}}{2}\right)\log\left\{\frac{a(\text{Mn}^{2+})\{a(\text{H}_2\text{O})\}^2}{a(\text{MnO}_2)\{a(\text{H}^+)\}^4}\right\}$$

Since $a(H_2O) = 1$ and $a(MnO_2) = 1$, and $a(Mn^{2+}) = c(Mn^{2+})/c^{\ominus} = 10^{-8}$ (given—assuming ideal behaviour), this becomes:

$$E = 1.23\,\text{V} - \left(\frac{0.059\,2\,\text{V}}{2}\right)\log(10^{-8}) + \left(\frac{0.059\,2\,\text{V}}{2}\right)\log\{a(\text{H}^+)\}^4$$

$$= 1.23\,\text{V} - (-0.237\,\text{V}) + \left(\frac{4 \times 0.059\,2\,\text{V}}{2}\right)\log a(\text{H}^+)$$

$$= 1.467\,\text{V} - (2 \times 0.059\,2\,\text{V})\,\text{pH}$$

$$= 1.47\,\text{V at pH 0;}\quad 0.52\,\text{V at pH 8}$$

By comparison, for the oxygen electrode,

$$E = 1.23\,\text{V} - (0.059\,2\,\text{V})\,\text{pH}$$

$$= 1.23\,\text{V at pH 0;}\quad 0.76\,\text{V at pH 8}$$

Thus in acid solution (assuming pH 0), E for the oxygen electrode is *less* positive than that for the manganese couple, so equilibrium should lie to the left. By contrast, at pH 8, the corresponding values are 0.76 V and 0.52 V, respectively, so thermodynamics predicts that increasing pH should indeed push the system in equation 60 over to the right—in agreement with Le Chatelier's principle.

SAQ 16
(Objectives 8, 9, 10 and 11)

(a) There are two approaches here. The first, and simpler, is to use the potential diagram itself, just as you did for acid solutions in Block 1. Thus, concentrating on the lower left-hand cycle in Figure 8b:

$$1 \times E^{\ominus}(\text{MnO}_4^- \mid \text{MnO}_4^{2-}) + 2 \times E^{\ominus}(\text{MnO}_4^{2-} \mid \text{MnO}_2) = 3 \times E^{\ominus}(\text{MnO}_4^- \mid \text{MnO}_2)$$

So $2E^{\ominus}(MnO_4^{2-} \mid MnO_2) = 3 \times (0.59\,V) - 0.56\,V$

and $E^{\ominus}(MnO_4^{2-} \mid MnO_2) = 0.61\,V$

Alternatively, the Nernst equation can be applied to the MnO_4^{2-}/MnO_2 couple, which is written for acid solutions (see the S343 *Data Book*) as:

$$\text{MnO}_4^{2-}(\text{aq}) + 4\text{H}^+(\text{aq}) + 2e = \text{MnO}_2(\text{s}) + 2\text{H}_2\text{O(l)};\quad E^{\ominus} = 2.27\,\text{V}$$

$$E = 2.27\,\text{V} - \left(\frac{0.059\,2\,\text{V}}{2}\right)\log\left\{\frac{a(\text{MnO}_2)\{a(\text{H}_2\text{O})\}^2}{a(\text{MnO}_4^{2-})\{a(\text{H}^+)\}^4}\right\}$$

with $a(MnO_2) = a(H_2O) = a(MnO_4^{2-}) = 1$, this becomes:

$$E = 2.27\,\text{V} - (2 \times 0.059\,2\,\text{V})\,\text{pH}$$

$$= 0.61\,\text{V at pH 14 (i.e. } a(\text{OH}^-) = 1)$$

(b) (i) Consistent. $E(MnO_4^- \mid MnO_4^{2-})$ is independent of pH, but, as you found in answering part (a), $E(MnO_4^{2-} \mid MnO_2)$ is strongly dependent. At pH 0, $E^{\ominus}(MnO_4^{2-} \mid MnO_2)$ is much greater than $E^{\ominus}(MnO_4^- \mid MnO_4^{2-})$, so manganese(VI) is thermodynamically unstable to disproportionation in acid solution. At pH 14, $E(MnO_4^{2-} \mid MnO_2)$ has dropped to 0.61 V, but is still just greater than $E^{\ominus}(MnO_4^- \mid MnO_4^{2-})$, so manganese(VI) is now *just* unstable. However, more alkaline conditions will lower $E(MnO_4^{2-} \mid MnO_2)$ still further, pushing it below the value for the $(MnO_4^- \mid MnO_4^{2-})$ couple and hence stabilising manganese(VI) to disproportionation.

(ii) At pH 0, the potential of the oxygen electrode is 1.23 V. As $E^{\ominus}(Mn^{3+} \mid Mn^{2+}) = 1.6\,V$ from Figure 8a, aerial oxidation of $Mn^{2+}(aq)$ to $Mn^{3+}(aq)$ cannot occur under these conditions.

At pH 14, the potential of the oxygen electrode is 0.40 V, and, according to the potential diagram in Figure 8b, $E^{\ominus}(MnO(OH) \mid Mn(OH)_2)$ is only $-0.29\,V$, so at this pH oxygen is thermodynamically capable of oxidising $Mn(OH)_2$ to the hydrated manganese(III) oxide, $MnO(OH)$.

SAQ 17
(Objectives 9 and 12)

There are several points to note:

1 The $E^{\ominus}$ values in Table 9 run in the sequence: bipy $\approx$ phen $>$ H_2O $>$ CN^-, whereas simple dominance of the LFSE part of $\Delta H_L^{\ominus}(Fe^{3+}, g)$ would argue for the sequence: $H_2O >$ bipy $\approx$ phen $> CN^-$. Taking the aquo couple as our base-line, CN^- can be said to stabilise the higher oxidation state (in accord with 'normal' behaviour), whereas phen and bipy appear to stabilise the $+2$ state (to a very similar degree).

2 One of the factors at work here is undoubtedly the fact that iron(II) and iron(III) have the configurations d^6 and d^5, respectively. Thus, for low-spin complexes, it is now the *lower* oxidation state where the contribution from the LFSE increases with Δ_o by the maximum amount ($\frac{12}{5}\Delta_o$). On the other hand, the contribution to $\Delta H_L^{\ominus}(Fe^{3+}, g)$—which is $2\Delta_o$ for d^5—is not very different; moreover, the actual *value* of Δ_o will be greater in the +3 state. The situation is far less clear cut than the Co^{III}/Co^{II} example in the text.

3 A further twist is introduced by the other thermodynamic data in Table 9. Thus, the values of $\Delta H_m^{\ominus}$(comp) are, in fact, *all more positive* than the value for the aquo couple, indicating that, on *energetic grounds alone*, CN^- (like phen and bipy) also stabilises the +2 state (albeit to a degree that does not reflect their relative positions in the spectrochemical series). So this is a case where the observed behaviour appears to turn on the contribution from the entropy term. Unfortunately, a detailed (and reliable!) interpretation of the latter quantity is even more difficult, especially for solution reactions involving ionic species.

ANSWERS TO EXERCISES

Exercise 1 (*revision*)

(a) Given the hint in the question, your sketch should have the same general features as the plot of $L_0(MCl_2, s)$ across the first transition series that you met in Block 2. Figure 12 is a plot of the actual values.

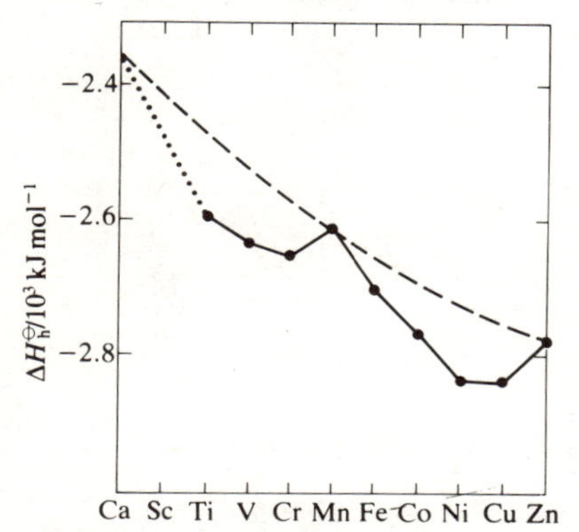

Figure 12 Variation in $\Delta H_h^{\ominus}(M^{2+}, g)$ across the first transition series.

The smooth curve (shown dashed) links the d^0, d^5 and d^{10} ions (Ca^{2+}, Mn^{2+} and Zn^{2+}) for which the LFSEs are zero. Hence it represents the variation in $\Delta H_h^{\ominus}$ that would be expected if *none* of the ions were stabilised by the ligand-field splitting of the 3d orbitals. Under these circumstances a simple electrostatic model will suffice. As we argued earlier in the Course, for successive ions one proton is added to the nucleus and one electron to the 3d shell. The increasing attractive force caused by the increasing nuclear charge outweighs the rise in the electron–electron repulsion, the observed result being a *decrease* in the size of M^{2+} across the series. Consequently, $\Delta H_h^{\ominus}$, the energy of interaction between the M^{2+} ion and the dipoles of surrounding water molecules, becomes steadily more negative along the dashed curve as the size of M^{2+} decreases.

Superimposed on this smooth decrease are the LFSEs typical of high-spin octahedral complexes, running from zero at d^0, to a maximum at d^3 (V^{2+}) and back to zero at d^5 (Mn^{2+})—with a repeat performance in the second half of the series. The net effect is to produce the characteristic double-bowl-shaped variation shown in Figure 12.

(b) $\Delta H_h^{\ominus}(M^{3+}, g)$ should again show the characteristic double-bowl-shaped variation described above. The most obvious difference is that the ions with the d^0, d^5 and d^{10} configurations are now Sc^{3+}, Fe^{3+} and Ga^{3+}: hence the double bowl should be slung below the values of $\Delta H_h^{\ominus}(M^{3+}, g)$ for these three ions. You may also have predicted two further features: first, a more marked *overall* decrease in $\Delta H_h^{\ominus}(M^{3+}, g)$ than in $\Delta H_h^{\ominus}(M^{2+}, g)$ because of the higher charge on the central ion; second, that the bowls will be deeper because the ligand-field splitting Δ_o is larger for the higher oxidation state.

As you will see, all three expectations are borne out in practice.

Exercise 2 (*Objective 2*)

(a) The completed cycle is shown in Figure 13. Here L_2 and L_3 represent the lattice energies of the dihalide and trihalide, respectively, which are virtually equal to $\Delta H_m^{\ominus}$ for the reactions:

$$M^{2+}(g) + 2Cl^-(g) = MCl_2(s)$$

and $$M^{3+}(g) + 3Cl^-(g) = MCl_3(s)$$

From the cycle:

$$\Delta H_m^{\ominus} = -L_2 + \Delta H_f^{\ominus}(Cl^-, g) + I_3 + L_3 \qquad (74)$$

As $\Delta G_m^{\ominus}(13) = \Delta H_m^{\ominus}(13) - T\Delta S_m^{\ominus}(13)$, and as $\Delta H_f^{\ominus}(Cl^-, g)$ is constant and $T\Delta S_m^{\ominus}(13)$ can be assumed to be nearly constant from metal to metal across the series:

$$\Delta G_m^{\ominus}(13) \approx I_3 - L_2 + L_3 + \text{constant} \qquad (75)$$

$$MCl_2(s) + \tfrac{1}{2}Cl_2(g) \xrightarrow{\Delta H_m^{\ominus}(13)} MCl_3(s) \qquad (13)$$

$MCl_2(s) \xrightarrow{-L_2} M^{2+}(g) + 2Cl^-(g)$; $\tfrac{1}{2}Cl_2(g) \xrightarrow{\Delta H_f^{\ominus}(Cl^-, g)} Cl^-(g)$; $M^{3+}(g) + 3Cl^-(g) \xrightarrow{L_3} MCl_3(s)$

$$M^{2+}(g) + 2Cl^-(g) + Cl^-(g) \xrightarrow{I_3} M^{3+}(g) + 3Cl^-(g)$$

Figure 13 Completed version of Figure 7.

(b) $\Delta G_m^{\ominus}$ and I_3 appear with the same sign on opposite sides of equation 75. Parallelism between $\Delta G_m^{\ominus}$ and I_3 implies that the variation in I_3 must be more significant than that in $(-L_2 + L_3)$. However, the latter variation is enough to render the parallelism imperfect. As for the aqueous ions, irregularities in the separate L_2 and L_3 variations are caused by ligand-field effects. These irregularities are the usual double-bowl shape, because all known dihalides and trihalides of the first-row transition elements are high spin and have structures in which the metal is octahedrally coordinated.

ACKNOWLEDGEMENTS

Grateful acknowledgement is made to the following source for material used in this Block:

Figure 3: D. A. Johnson (1982, 2nd edn) *Some Thermodynamic Aspects of Inorganic Chemistry*, Cambridge University Press.

S343 Inorganic Chemistry